UNITEXT for Physics

More information about this series at http://www.springer.com/series/13351

Rosa Poggiani

Optical, Infrared and Radio Astronomy

From Techniques to Observation

 Springer

Rosa Poggiani
Department of Physics
University of Pisa
Pisa
Italy

ISSN 2198-7882   ISSN 2198-7890 (electronic)
UNITEXT for Physics
ISBN 978-3-319-83123-7   ISBN 978-3-319-44732-2 (eBook)
DOI 10.1007/978-3-319-44732-2

Printed on acid-free paper

This Springer imprint is published by Springer Nature
The registered company is Springer International Publishing AG
The registered company address is: Gewerbestrasse 11, 6330 Cham, Switzerland

In loving memory of my mother Anna, an inquiring mind and a courageous heart

Preface

Optical, infrared, and radio astronomies are the historical pillars of astrophysics. They are continuing to contribute a large part of the astrophysical information and are the prerequisite for observations with highenergy information carrier. The investigation of the astrophysical properties of the targets triggers the building of the large observational facilities and shapes their characteristics. The textbook is organized from the point of view of the science targets, tackling optical, infrared, and radio astronomies as scientific research areas. In place of presenting the observational techniques and showing how they could be used in different domains of the electromagnetic spectrum, the textbook is focused on the science targets and the measurement of their fluxes and spectra, providing a link between observational techniques and astrophysical science.

The textbook has grown up out of several years of teaching the courses of astrophysical techniques to graduate students at the University of Pisa who were specializing in astrophysics. The text shows the state of the art and the future evolution of instrumentation and observational methods. The aim of the work is to be a comprehensive guide through the steps needed to acquire and analyze optical, infrared, and radio data: planning the observation, choosing the signal-to-noise ratio, selecting a telescope or radio telescope with the suitable instrumentation to observe the selected object at the proper epoch, performing the observations, securing the calibration data, and extracting the astrophysical information as fluxes or spectra. Thus, for each astronomy research area investigated in the book, the relevant orders of magnitudes are firstly presented. Then, the physical principles of the telescopes, the detectors, and the components needed for flux and spectra measurements are discussed. The signal-to-noise ratio of the observations and the limits of instrumentation are discussed in detail in view of writing proposals for telescope observing time. Finally, the data analysis techniques are presented. The bibliography at the end of each chapter suggests monographs of interest for the reader. Web links are provided for the instrumentation.

The first part of the textbook is devoted to the basics of astronomical observations: the electromagnetic radiation and its interactions, the effect of the atmosphere on observations, and the observational windows (Chap. 1). Then, the ingredients needed to point an instrument for observations are presented: the celestial coordinates, the measure of time, the astronomical nomenclature, and the source catalogs, the steadily increasing Internet resources available for observers (Chap. 2). The second part of the text is devoted to the techniques of optical astronomy. Firstly, the optical telescopes and the involved aberrations are discussed, together with the configurations of reflector instruments that are the core of large observing facilities (Chap. 3). The ground-based optical telescopes are later discussed, investigating the effects of the atmosphere on the quality of images and discussing the operation of telescopes in space (Chap. 4). The radiation collected by optical telescopes is measured by light detectors; among them, the charge-coupled devices that have revolutionized observational astronomy are discussed in detail (Chap. 5). The measurement of the optical fluxes by the technique of optical photometry is discussed in Chap. 6, with a focus on photometry with CCDs. The measurement of optical spectra, mainly using dispersing elements, is discussed in Chap. 7. The third part of the book is devoted to the low-energy side of classical astronomy. Infrared astronomy, the domain of thermal emission, is presented, with a discussion of the specific observational techniques in the different wavelength regions (Chap. 8). The techniques and the instruments for data collection in radio astronomy, the first astronomy after the optical one, are discussed later (Chaps. 9 and 10). The fourth part of the book deals with the combination of optical telescopes and radio telescopes in interferometric arrays to achieve a high angular resolution (Chaps. 11 and 12). The fifth part of the text is devoted to present the preparation and execution of observations (Chap. 13) and the data analysis techniques (Chap. 14).

The author is greatly indebted to several people. I am very grateful to my friends and colleagues Dario Grasso, Scilla Degl'Innocenti, Ivan Bruni, Valentina Cettolo, Franco Giovannelli, Andrea Macchi, Antonio Marinelli, and Ignazio Bombaci, and to my advisor and mentor Gabriele Torelli, for the interesting discussions about physics and astrophysics and their constant support. Many thanks to the Time Allocation Committee and the staff of the Loiano Observatory for the observation time. I am very grateful to my friends Rita Mariotti and Paolo Pancani for their support. I thank the students who attended my courses at the Department of Physics of University of Pisa, for their interest and their questions. Thanks to the technicians of the student laboratories. I am indebted with Barbara Amorese and Marina Forlizzi at Springer, for their suggestions and the professional and kind support.

Last but not least, the author thanks her mother Anna who shared the dream of the book, but could not see it in print. Her lifelong support and encouragement have made this book a reality.

Pisa, Italy   Rosa Poggiani
July 2016

Contents

Part I
The Basics

Chapter 1
Setting the Scene

This chapter presents the main properties of the electromagnetic spectrum, the main information carrier in astronomy. The optical, radio, and infrared astronomies, on the lower energy side of the spectrum, are the subjects of this book. The main properties of the electromagnetic radiation are reviewed, discussing the physical variables that describe the energy transported by the radiation. The original information of the astrophysical source is modified during the travel to the observer, by the interstellar medium and by the Earth atmosphere. The astronomy has been restricted to the window of visible light until the advent of radio astronomy. The atmosphere has an high absorbing power in the other regions of the electromagnetic spectrum. The advent of observatories in space has allowed the access to the whole electromagnetic spectrum. The scattering of light prevents observations during the daytime. In addition, the thermal emission of the atmosphere is the source of a large background to the infrared observations.

1.1 Astronomy as an Observational Science

Astronomy is an ancient science and has triggered the development of other sciences. But it is completely different from experimental sciences: astronomy is an observational science. The subjects of the research, the celestial bodies, are neither accessible nor easily reproducible in the laboratory. Several systems show periodic brightness variations, but the events of interest for the astronomers are generally unpredictable. There is no control over the observing conditions. In addition, the information secured by astronomers is degraded by the media that the radiation has encountered during the travel to the observer. The sources of background and noise are very relevant to astronomy: generally, the signal-to-noise ratio is low and long observation times are needed. We will present the fundamental backgrounds to observations in this chapter. The noise sources related to the instrumentation are linked

© Springer International Publishing Switzerland 2017
R. Poggiani, *Optical, Infrared and Radio Astronomy*,
UNITEXT for Physics, DOI 10.1007/978-3-319-44732-2_1

to the physical process used for detection and will be discussed in the following chapters.

Given the low signal-to-noise ratios generally encountered in astrophysics, a large part of observing time is spent in securing calibration data, observing celestial sources whose flux and spectra have been measured with great precision. Due to the variability of factors affecting an astronomical observation, calibration data must be measured for each session. To date, the archives of large telescopes or satellite missions are publicly available: the advent of Internet has made astronomical research possible without observing in person. Care must be taken to check the observing conditions of the data.

We may wonder now which are the mechanisms of astronomical discovery: is it driven by theoretical predictions or by observations? Probably, it is a mixture of both. General Relativity has been confirmed by the observation of the solar eclipse in 1919. Neptune has been discovered on the basis of prediction based on the anomaly of Uranus orbit. However, serendipity seems to be the rule in astronomy. Pulsars were discovered during a systematic survey in the radio domain. Gamma-ray bursts were discovered by satellites looking for clandestine nuclear tests. As in other sciences, technology has a strong role. The great leaps in astronomy are historically linked to the advent of innovative instrumentation. The advent of the telescope at the beginning of the seventeenth century has opened a new way to observe the universe. The spectroscopic instrumentation has shown the composition of celestial objects, made of the same elements found on Earth. The invention of photographic plates has given the first recording system independent from the observer. The advent of charge-coupled devices has made possible to work with digitized data. The satellites have opened new observational windows. The progress of optics has produced telescopes with larger and larger apertures that combined with the improvement of the detector efficiency is continuing to change our view of the universe.

1.2 The Electromagnetic Spectrum: The Low-Energy Side

Modern astronomy is intrinsically multimessenger, since it uses a great variety of probes: the whole electromagnetic spectrum, cosmic rays, neutrinos, and gravitational wave. Narrowing the discussion to the electromagnetic spectrum, we deal with multiwavelength astronomy. The aim of observational astronomy is the collection of the energy from the whole spectral energy distribution of sources. Traditionally, the different domains of electromagnetic spectrum are classified using wavelength units above 1 nm and energy units below, starting from about 100 eV. The classification will be used to present the different astronomies in the following chapters:

- Gamma-rays: energy above 1 MeV;
- X-rays: energy in the range 1 keV–1 MeV;
- Ultraviolet (UV): wavelengths in the range 10–300 nm;
- Optical: wavelengths in the range 300–800 nm;

- Infrared (IR): wavelengths in the range 0.8–200 micron;
- Submillimeter: wavelengths in the range 0.2–1 mm; and
- Radio: wavelengths above 1 mm.

The reconstruction of spectral energy distribution of a celestial object requires the combination of the observations of different instruments operating in the different domains of the spectrum, not always possible. The response of instrumentation is critically dependent on the energy range and, on Earth, on the transmission properties of the atmosphere. Each astronomical observation intercepts a slice of the spectral distribution in a range of wavelengths, selected by the instrumentation. This is the reason why we are dealing with different astronomies: optical, infrared, ultraviolet, and radio, with X-rays or γ-rays.

The low-energy side of the electromagnetic spectrum, spanning the optical, infrared, and radio astronomies, is the subject of the present book. The instrumentation and the observational techniques in these domains are very well developed and provide a large part of the astronomical information, also in support of observations at high energies. The properties of the incident electromagnetic radiation can be described by modeling it as an energy flow along some direction [4]. The *specific intensity* or *specific brightness* is the rate of power P transported along a chosen direction per unit of area A, of solid angle Ω, of frequency ν:

$$I_\nu = \frac{dP}{dA\, d\Omega\, d\nu} \tag{1.1}$$

The product $dA\, d\nu$ is called *etendue* or *throughput*: it is a conserved quantity in ideal optical systems. The *flux density* is the energy flux across an area A:

$$F_\nu = \int I_\nu \cos\theta d\Omega \tag{1.2}$$

where θ is the angle between the direction of radiation and the normal to the area. A suitable approximation for astronomical sources with uniform brightness is $F_\nu = I_\nu \Omega_s$, where Ω_s is the solid angle subtended by the source. The specific intensity and the flux density can be defined per unit wavelength as:

$$I_\lambda = \frac{dP}{dA\, d\Omega\, d\lambda} \tag{1.3}$$

$$F_\lambda = \int I_\lambda \cos\theta d\Omega \tag{1.4}$$

The two definitions of specific intensity are related by $\nu F_\nu = \lambda F_\lambda$.

Due to the faintness of astronomical sources, the astronomical flux unit is the Jansky (Jy):

$$1 Jy = 10^{-26} \frac{W}{m^2\, Hz} \tag{1.5}$$

The spectrum of many astrophysical sources can be approximated by a black body [4] distribution. The specific intensity of a black body at temperature T is given by the Planck distribution:

$$B_\nu(T) = \frac{2h}{c^2} \frac{\nu^3}{e^{\frac{h\nu}{kT}} - 1} \tag{1.6}$$

The position of the maximum of the black body distribution is described by the *Wien law*, $\lambda_{max}(\mu m)T(\mathrm{K}) = 2898$. The limits of the black body distribution at high and low frequencies are:

- Wien limit, $h\nu \gg kT$:

$$B_\nu(T) = \frac{2h\nu^3}{c^2} e^{-\frac{h\nu}{kT}} \tag{1.7}$$

- Rayleigh–Jeans limit, $h\nu \ll kT$, typical of radio astronomy:

$$B_\nu(T) = \frac{2kT}{c^2}\nu^2 \tag{1.8}$$

As an example, the spectrum of the Sun is reported in Fig. 1.1 together with the normalized spectrum of a black body at 6000 K.

Photons in thermal equilibrium at temperature T follow the Planck distribution. The mean number of photons is:

$$\bar{n} = \frac{1}{e^{\frac{h\nu}{kT}} - 1} \tag{1.9}$$

with a mean energy:

$$\bar{E} = \frac{h\nu}{e^{\frac{h\nu}{kT}} - 1} \tag{1.10}$$

Fig. 1.1 Spectrum of the Sun compared with the normalized distribution of a black body at 6000 K

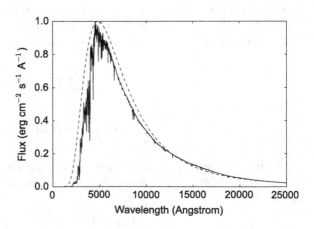

The mean square fluctuation of the number of photons at temperature T is:

$$(\Delta n)^2 = \bar{n} + \bar{n}^2 \tag{1.11}$$

In the Wien limit discussed above, the fluctuation of the number of photons is $\sim \bar{n}$, the familiar result of the Poisson distribution, that deals with uncorrelated events with random arrival times. In the Rayleigh–Jeans limit, the fluctuation becomes $\sim \bar{n}^2$, deviating from Poisson statistics: photons are correlated. The two limits define the regimes where the photons can be treated as particles or classical electromagnetic waves. The optical astronomy domain belongs to the former regime, while the radio astronomy to the latter. The detection in the optical region is incoherent, while the coherent detection of the electric field of the incident wave is possible in the radio domain. Infrared and submillimeter detection techniques are at the border.

Optical astronomy has devised the historical *magnitude* system as a relative measure of the flux of a source [4]. The origins of the magnitude system date back to the ancient visual observations by the Greek astronomer Hipparcos. The stars were divided into six groups separated by equal steps. The brightest stars were classified in the first magnitude group, while the stars barely visible with the naked eye were classified in the sixth magnitude group. In 1856, Pogson proposed the logarithmic relationship between the visual magnitude and the logarithm of the star intensity. Since the magnitude is a relative measure of the flux, it is necessary to define a reference source. The *monochromatic magnitude* m_{λ_0} at the wavelength λ_0 is defined starting from the monochromatic flux $F(\lambda_0)$ and a reference flux F_0:

$$m_{\lambda_0} = -2.5 \log \frac{F(\lambda_0)}{F_0} = -2.5 \log F(\lambda_0) + m_0 \tag{1.12}$$

The magnitude scale shows some counterintuitive aspects. It is a reversed scale, where brighter objects are assigned smaller magnitudes than fainter objects, and is a logarithmic scale. The presence of a reference flux introduces a constant term m_0 in magnitude, the *zero point*. For two objects of magnitude m_1 and m_2 with fluxes F_1 and F_2, the difference in magnitude is:

$$m_1 - m_2 = -2.5 \log \frac{F_1}{F_2} \tag{1.13}$$

Photometric observations are secured with systems of filters defined by transmission curves. The magnitude in the band T defined by the filter with transmission $T(\lambda)$ is:

$$m_T = -2.5 \log \int T(\lambda)\, F(\lambda)\, d\lambda + m_{T,0} \tag{1.14}$$

where $m_{T,0}$ is the zero point of the photometric band. The details of the photometric systems are described in Chap. 6.

The *bolometric magnitude* is the integral of the spectral distribution over the whole electromagnetic spectrum:

$$m_{bol} = -2.5 \log \left(\int_0^\infty F(\lambda) \, d\lambda \right) + constant \qquad (1.15)$$

The estimation of the energy distribution over the whole spectrum requires the combination of the data of instruments operating over different spectral regions.

The flux of an observed astronomical source depends on its intrinsic brightness, but also on the distance. The flux collected by the instrumentation defines the *apparent magnitude m*. The *absolute magnitude M* is the magnitude that the source would show if placed at the standard distance of 10 parsec. The apparent and absolute magnitude are related by:

$$m - M = 5 \, \log \frac{d}{10 \, pc} = 5 \log d - 5 \qquad (1.16)$$

The quantity $m - M$ is called the *distance modulus*.

The absolute bolometric magnitude M_{bol} is related to the luminosity [1]:

$$L = 3 \times 10^{28} \times 10^{-0.4 M_{bol}} \, W \qquad (1.17)$$

1.3 Intergalactic and Interstellar Media

The radiation encounters different media during the trip to Earth, the intergalactic and interstellar media. The interstellar medium consists of dust and gases and produces absorption and scattering of photons. The effect is taken into account including an additional term A, the *extinction*, to the right-hand side of Eq. 1.16. The visual extinction is of the order of a few magnitudes per kpc above the galactic plane, but is about 30 magnitudes in the direction of the galactic center, that is investigated in the infrared. The effects of the interstellar reddening on astronomical photometry are discussed in Chap. 6.

The interstellar medium is ionized, thus the radio waves undergo dispersion and Faraday rotation. The dispersion of the waves produces a delay in the arrival times of waves with different frequencies, proportional to the *dispersion measure*:

$$DM = \int N dl \qquad (1.18)$$

where N is the electron density of the interstellar medium and l the distance of the source. The Faraday rotation is the rotation of the polarization angle of the radiation produced by the magnetic field B of the interstellar medium. The effect is proportional to the *rotation measure*:

$$RM = \int N B dl \qquad (1.19)$$

1.4 The Atmosphere: Absorption, Scattering, and Emission

The atmosphere is opaque or partially opaque to radiation belonging to different parts of the electromagnetic spectrum [4]. To date, more and more observatories operate in space. Only through observations from satellites, the whole electromagnetic spectrum has become accessible to astronomy. Even in the spectral regions where ground-based observations are possible, space observatories produce higher quality data. Observations from ground are still very popular. Access is easier, the cost to build instrumentation is smaller, and larger telescope apertures can be realized. The penalty is the presence of the atmosphere that produces extinction of radiation and a deterioration of the image quality. The effect can be mitigated by the use of adaptive optics. The operation of telescopes in space allows the access to additional observational windows beyond the optical and radio ones. Two technical considerations split the wolds of ground-based and space-based observations: a difference in cost by one or two orders of magnitude and the difference in the lifetime of a facility that drops from the tens years of ground-based facilities to a few years for space observatories.

The behavior of the temperature versus the altitude for the standard atmosphere defined in 1976 [7] is reported in Fig. 1.2.

The temperature shows an alternation of increasing and decreasing trends in the troposphere, stratosphere, and mesosphere that are separated by inversion layers (tropopause and stratopause). Considering an isothermal atmosphere with an average temperature T_0, the pressure P decreases as:

Fig. 1.2 The US standard atmosphere: behavior of the temperature as a function of altitude (data available at http://www.pdas.com/atmos.html)

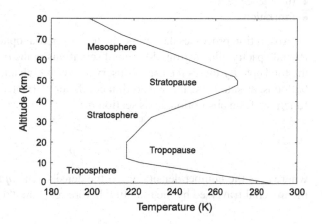

$$P(z) = P_0 \exp\left(-\frac{x}{H_0}\right), \tag{1.20}$$

where $H_0 = \frac{RT_0}{Mg}$ is the scale height, R the gas constant, and M the average molecular mass of atmosphere; since nitrogen is the main constituent, the scale height is of the order of 8 km. The atmosphere shows instabilities governed by the adiabatic gradient. Whenever the gradient in temperature exceeds the adiabatic gradient, convective instabilities, vertical motions of air masses, will occur. The composition of atmosphere includes not only nitrogen and oxygen, but also several minority components that have an impact on astronomical observations [4]. The water vapor plays a strong role. The distribution of water vapor has a typical scale height of about 3 km, smaller than the scale height of the isothermal atmosphere. As a consequence, astronomical observatories must be built at an altitude of a few kilometers, typically on mountains. The water vapor is particularly harmful for infrared observations, causing total opacity in some spectral regions and leaving a few available slots for ground-based observations. The infrared astronomy is affected also by another minor atmospheric component, the carbon dioxide (CO_2), that is strongly absorbing. The concentration of CO_2 has been steadily increasing in the last decades. The ozone O_3 is produced by the interactions of the ultraviolet solar radiation with the atmosphere in the altitude interval from 15 to about 35 km. Ozone is the responsible for the transparency cutoff of radiation at about 300 nm, forcing ultraviolet observation to be performed from space.

The physical processes contributing to the opacity of the atmosphere and to the background of astronomical observations are:

- absorption,
- emission,
- scattering,
- refraction and dispersion,
- turbulence, and
- ionization.

Absorption processes determine the total or partial opacity of the atmosphere [4]. Partial opacity allows ground-based observations in the *observational windows*, but the total opacity forces the use of observatories at high altitudes, on board of planes, balloons, or satellites. The absorption coefficient k related to a physical process is a function of the absorption cross section σ_i:

$$k_i = \frac{\sigma_i n_i}{\rho} \tag{1.21}$$

where n_i is the number density and ρ the density. The *optical depth* τ produced by the material traversed by the radiation before reaching the observer is defined as:

$$\tau = \int_{h_1}^{h_2} \sum_i k_i \rho \, dh, \tag{1.22}$$

where h_1 is the altitude of the observatory, h_2 the altitude of the top of the atmosphere, and the sum is performed over all the atmospheric components. The amount of transmitted radiation, for an initial intensity I_0, behaves as a decreasing exponential:

$$I = I_0 \exp\left(-\tau \frac{1}{\cos z}\right), \tag{1.23}$$

where the factor $\frac{1}{\cos z}$ accounts for a generic incidence angle z. For convention, the angle is measured from the zenith direction and is called *zenith angle*. The altitude at which the observatory must be placed depends on the fraction of the residual incident flux that is considered acceptable. Usually, the altitude is chosen where the flux of incident radiation is reduced by 50%, that corresponds to an optical thickness of about 0.69.

The absorption processes involve rotational, roto-vibrational, or electronic transitions of the main atmospheric components: N_2, O_2, O_3, H_2O, CO_2, atomic nitrogen, and oxygen. The absorption in the UV is produced by atomic and molecular interactions with oxygen and nitrogen. The infrared absorption is determined by rotational and roto-vibrational transitions of water vapor and carbon dioxide. The attenuation in the radio region is produced by the interactions of the radio waves with the electron and ions produced by the photochemical reactions of the solar radiation with the atmospheric molecules in the ionosphere, an atmospheric layer extending from 60 to about 300 km altitude. The refraction index of the ionized layers is:

$$n = \sqrt{1 - \frac{\omega_p^2}{\omega^2}} \tag{1.24}$$

where ω_p is the plasma frequency:

$$\omega_p = 2\pi \left(\frac{n_e e^2}{4\pi^2 \varepsilon_0 m_e}\right)^{\frac{1}{2}} \tag{1.25}$$

where m_e is the electron mass and n_e the electron number density. Electromagnetic waves with a frequency smaller than plasma frequency will be attenuated. For the typical value of electron density of about 10^6 cm^{-3}, the cutoff is at a frequency of a few MHz, i.e., at wavelengths above a few meters. Thus, the centimeter radio waves can be detected by ground-based instrumentation.

The emission of the atmosphere is the source of a relevant background in the low-energy part of the electromagnetic spectrum [4]. The emission has two components, the *fluorescence* and the *thermal emission*. The fluorescence emission, called *airglow*, is caused by the recombination of electrons and ions produced by photodissociation in the ionospheric layers. The airglow shows fast variations at the scale of a few minutes. The thermal emission of the atmosphere can be modeled considering the atmosphere as a black body with a mean temperature of 250 K; the maximum of emission is in the infrared region.

The incident radiation undergoes scattering with the atmospheric gas molecules and the aerosols (pollution, volcanic ashes, pollen, etc.) [4]. The diffusion of radiation on air molecules in the optical and infrared domains is governed by the *Rayleigh scattering*, whose cross section is:

$$\sigma_{Rayleigh} = \frac{8\pi^3}{3} \frac{(n-1)^2}{\lambda^4 N^2}. \tag{1.26}$$

where n is the refraction index of the atmosphere and N the number density of molecules. The Rayleigh scattering is responsible for the blue color of the sky during daytime and the faint light level during the night. The scattering on the aerosols occurs through the *Mie scattering*, since the scattering centers are larger than the air molecules. The relevant property of the Mie scattering cross section is the dependence on the wavelength, $\sigma_{Mie} \propto \lambda^{-1}$, less steep than the Rayleigh one $\sigma_{Rayleigh} \propto \lambda^{-4}$. The contribution of aerosols is very variable and depends on weather, pollution, volcanic events, etc.

The incident radiation undergoes refraction and dispersion in the atmosphere [4]. The refraction is the reason of the flattening of the Sun disk shape close to the horizon. The index of refraction n of dry air at the standard temperature and pressure is given by [1]:

$$(n-1) \times 10^6 = 64.328 + \frac{29498.1}{146 - \lambda^{-2}} + \frac{255.4}{41 - \lambda^{-2}} \tag{1.27}$$

where the wavelength is measured in μm. Refraction produces an apparent shift of the star position toward the zenith that increases with increasing zenith distance. Due to the refraction, the Sun is still visible after sunset when it is slightly below the horizon. The refraction is a chromatic effect, as shown by the occurrence of the green flash at sunset and sunrise. It produces a different shift for stars with different colors, and a rainbow-like spread of the image of stars observed at large zenith angles. The dispersion must be compensated in high-resolution astrometric and interferometric observations.

The atmosphere is not a static medium, but its properties are changing in time triggered by several effects. The first mechanism is the *convection* that occurs when the temperature gradient exceeds the adiabatic gradient. Another mechanism is the occurrence of turbulence [4]. The *Reynolds number* of a fluid is defined as:

$$Re = \frac{Lv}{v_c} \tag{1.28}$$

where L is the scale of irregularities in the medium, v the fluid velocity, and v_c its kinetic viscosity. Turbulence develops when the Reynolds number becomes larger than a threshold value, about 2000 [4]. Assuming $L \sim 10$ m, $v \sim 5$ m s^{-1}, and $v_c \sim 1.5 \times 10^{-5}$ m^2 s^{-1}, the Reynolds number is about 3×10^6, much larger than the critical value. The details of the effects of turbulence in astronomical observations will be presented in Chap. 4.

1.5 Observational Windows

The existence of *observational windows* for ground-based observations is a consequence of the physical processes described above. For completeness, we report in Fig. 1.3 the altitude and the fraction of atmosphere at which one half of the incident radiation is transmitted for the whole electromagnetic spectrum [4, 5].

Ground-based astronomy can be performed only in some parts of the electromagnetic spectrum, in addition to the visible window. Radio astronomy, the first historical astronomy after the optical one, is feasible from ground only for wavelengths above a few centimeters. The close region of the infrared (below about 25 μm) and the millimeter region are accessible from ground, with the exceptions of some opaque regions. The other parts of the electromagnetic spectrum must be explored by observatories at high altitude or in space. Infrared and submillimeter observations can be performed starting from the altitude of airplanes. The observations of high-energy astrophysics (ultraviolet, X-rays, and gamma-rays) demand observations in space. Space-based observatories produce high-quality data than terrestrial-based observatories. The present textbook describes the different solutions for ground-based and space-based observatories in the optical, infrared, and radio domains. The astronomies with X-rays, gamma-rays, cosmic rays, neutrinos, and gravitational waves are discussed by [6].

The atmospheric processes described in Sect. 1.4 govern the selection of a ground-based site for optical, infrared, and millimeter observatories [2] and allow to answer to a common question: why do astronomers select exotic locations for their telescopes? The first constraint for an observatory site is a minimal coverage by clouds. Clouds are

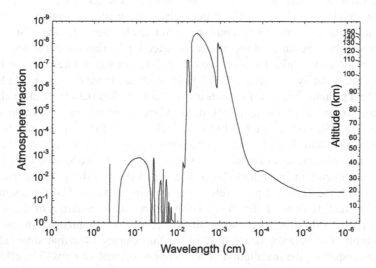

Fig. 1.3 Atmospheric absorption of photons in the Earth atmosphere. The abscissa reports the wavelength of the radiation. The *curve* is the altitude or atmospheric fraction at which one half of the incident radiation is transmitted. The figure has been adapted from [5]

stopped at low altitudes by temperature inversion layers. Large observing facilities are built on mountains. Observatory sites must also have a minimal amount of water vapor, especially for infrared and submillimetric observatories, where it produces a strong absorption. Sites in the tropical or desert regions are the best choices for optical and infrared observatories, with a mean column of precipitable water of the order of a few mm. The amount is smaller than 1 mm in Antarctica (at an average altitude of 3 km) that has a thermally stable atmosphere and a low local temperature, useful to reduce the infrared background radiation. The atmosphere at the observatory site should show a low turbulence level. Islands and coastline sites are both good choices for observatory sites. Mountains in islands offer a stable atmospheric environment and have been chosen for the observatories at Canary islands and at the Hawaii. Sites close to coasts benefit from the smaller altitude of the inversion layer due to the low temperature of the sea; this is the type of site chosen for Chile observatories. Finally, the atmospheric transparency must be stable over the night. Radio astronomy is less harmed by the atmosphere opacity, but it is strongly affected by the radio and microwave emission of a large variety of emitters: radars, cell phones, high voltage instrumentation, etc. Radio astronomy is performed in a set of protected bands, like the band around the 21 cm transition of hydrogen.

The strong effect on the atmosphere on the astronomical observations suggests to investigate the possibility of observations from space. At an altitude of a few hundreds kilometers, the absorption of a large part of atmosphere is removed. However, the sky above the atmosphere is not completely black. The Solar System contains the *Zodiacal Cloud*, a system of dust grains extending up to a few astronomical units from the Sun and lying within a few degrees from the ecliptic plane. The dust grains are heated by the Sun to temperatures as high as the room temperature on Earth, causing a relevant thermal emission. They also act as scattering centers for solar radiation, making the cloud visible at good observatory sites as zodiacal light. The optical and near-infrared backgrounds are dominated by the scattering of solar light on the dust, while the infrared region is dominated by the thermal emission.

The choice of the orbit for space-based observatories is strongly driven by the science targets and by cost issues [2, 4]. High orbits are more expensive, but have a smaller background. The *Low Inclination Low Earth Orbits* (LEO) are at an altitude that ranges from 300 km up to 1000 km, to reduce the atmospheric friction and to avoid the Van Allen belts. The Hubble Space Telescope (HST) is in a low orbit with an altitude of 600 km. The low orbits are very popular due to the lower cost for launch and the larger payload that can be sent in space. The STS Space Shuttle has released several instruments in space and has demonstrated the possibility of maintenance in space, with the Hubble Space Telescope servicing mission. Instrumentation on board of a satellite in a low Earth orbit experiences an occultation of the science targets by the Earth; thus, long observations require a new acquisition of the target at each orbit. The periodic occultation of the Sun causes a thermal stress at each transition, requiring the installation of temperature control systems. The effects of the Sun can be minimized by choosing *Sun Synchronous Orbits*, with an orbital plane fixed relatively to the Sun. The *High Circular Orbits* are at an altitude of 6000 to 100,000 km, to avoid the Van Allen belts. The high orbits offer a smaller occultation by the Earth, but they are more expensive that low orbits. The *Elliptical Orbits* are a compromise solution, since the instrumentation mostly operates at large distances from Earth and can use the passage close to Earth to transmit data. The five *Lagrange points* are locations where the combination of the gravitational potential of the Earth and the Sun has a minimum. The point L_2 lies along the Sun–Earth direction, at 1.5×10^6 km toward the exterior of the Solar System and has been selected for the WMAP, FIRST, GAIA, HERSCHEL, and JWST observatories.

The Moon could be a suitable solution for a space-based observatory [4]. The reduced gravity allows the use of light materials for construction of the facilities, but the thermal environment shows a large temperature variation (from 90 K during the lunar night to 400 K during the lunar day) that has an impact on the mechanical alignment of the optical components of the instrumentation.

Instrumentation operating in space is exposed to the environmental radiation, the charges trapped in the Van Allen belts, for lower orbits, and cosmic rays and products of solar flares. The radiation produces an increased background in the detectors and a degradation of their performances. The detector and the electronics must be designed with the techniques of the *radiation hardness* to minimize the effects on long exposures.

1.6 Backgrounds

The physical processes discussed above produce an irreducible background to the astronomical observations that depends on the site of the observatory [2–4]. The sources of the background have an astrophysical or atmospheric origin. The airglow emission and the scattering of light (including artificial lights) dominate the optical spectral region. The night sky brightness shows long-term variations related to solar activity and short-time variations during the same night. The optical sky spectrum measured at the Gemini Observatory is reported in Fig. 1.4. The spectrum shows a

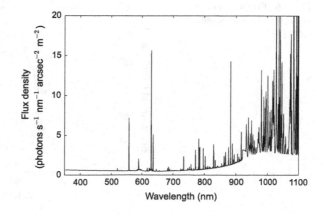

Fig. 1.4 Optical sky background at the Gemini Observatory, data available at http://www.gemini.edu/sciops/telescopes-and-sites/observing-condition-constraints/optical-sky-background

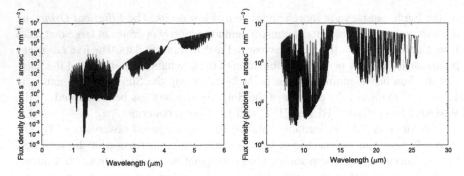

Fig. 1.5 Infrared sky background at the Gemini Observatory, data available at http://www.gemini.edu/sciops/telescopes-and-sites/observing-condition-constraints/ir-background-spectra

continuum and several emission lines superimposed: the lines of O I at 5577, 6300 Å, O_2 at 762 nm. The night sky spectra secured at observatories show also some artificial lines of sodium and mercury, caused by the lamps for street illumination.

The infrared background at the Gemini Observatory is reported in Fig. 1.5. The background emission from OH dominates the visible and the near-infrared region up to 2.3 μm. The OH emission is variable on the timescale of minutes. The thermal emission of the atmosphere and the telescope produces a strong contribution in the infrared region above about 2.3 μm.

The background in space-based observatories consists of several astrophysical contributions (Fig. 1.6) [3]. The contribution of zodiacal light shows the two components, produced by the scattering of the solar radiation and by the thermal emission from the dust. The scattered and the thermal spectra leave a region at about 3.5 μm where their contribution to the background is minimum, the *cosmological window*. The effect of the Zodiacal Cloud could be removed only by placing an observatory at several astronomical units from Earth. In addition, there are other astrophysical

Fig. 1.6 Brightness of the background sources, data from [3]

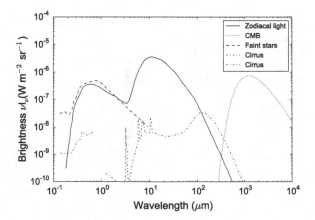

backgrounds, the light of unresolved faint stars, and the infrared cirrus. The radio region is dominated by the Cosmic Microwave Background (CMB).

Problems

1.1 Discuss the magnitude scale and the role of the zero point of the scale.

1.2 Discuss the main criteria for the choice of ground-based observatory sites for optical and infrared astronomy.

1.3 Discuss the reasons to build space-based observatories.

References

1. Allen, C. W. (Cox, A. N., Editor): Allen's Astrophysical Quantities. Springer-Verlag (2013)
2. Bely, P.: The Design and Construction of Large Optical Telescopes. Springer, New York (2003)
3. Leinert, Ch., et al.: The 1997 reference of diffuse night sky brightness. Astron. Astrophys. Suppl. Ser. **127**, 1 (1998)
4. Lèna, P., et al.: Observational Astrophysics. Springer-Verlag Berlin Heidelberg (2012)
5. Oda, M.: X-ray and-ray astronomy. Proc. ICRC **1**, 680 (1965)
6. Poggiani, R.: High Energy Astrophysical Techniques (2016), doi:10.1007/978-3-319-44729-2
7. U.S. Standard Atmosphere, 1976, prepared jointly by the National Oceanic and Atmospheric Administration, National Aeronautics and Space Administration and the U.S. Air Force, U.S. Government Printing Office, Washington, D.C. (1976)

Chapter 2
Pointing the Telescope: Astronomical Coordinates and Sky Catalogs

The coordinates of the objects in the sky are of paramount importance for observational astronomers. The position of known targets is necessary to plan the observations. The identification of a new source is always accompanied by the determination of the position with high precision and the comparison with catalogs of known objects. In the following, the coordinates and the measure of time in astronomy will be discussed with the practical purpose of planning an observation. A detailed technical discussion of astronomical coordinates can be found in [1, 2]. The evolution of Internet has made the planning of observations easier. The coordinates of an astronomical object can be accessed on line, building the charts of the sky region around the target and accessing the related literature. The sky surveys have produced a large number of catalogs of sources at all wavelengths.

2.1 Astronomical Coordinate Systems

The celestial objects appear to be on the *celestial sphere*, an arbitrarily large sphere enclosing the Earth. A system of *celestial* coordinates uses two spherical coordinates, similar to the terrestrial latitude and longitude [1–3]. A celestial coordinate system is defined by a fundamental circle and a fixed point on it. The terrestrial equator and the meridians are examples of *great circles*, circles on the sphere defined by a plane crossing the center of sphere; on the other hand, the parallels are examples of *small circles*, since their planes do not cross the Earth center. Some basic definitions will be used in the following:

- Celestial Poles: the North and South Celestial Poles are defined by the Earth axis, as the projections of the terrestrial poles on the celestial sphere; the elevation of the North Celestial Pole above the horizon is equal to the latitude of the observatory.
- Celestial Equator: the projection of the Earth equator on the celestial sphere.

© Springer International Publishing Switzerland 2017
R. Poggiani, *Optical, Infrared and Radio Astronomy*,
UNITEXT for Physics, DOI 10.1007/978-3-319-44732-2_2

- Zenith, Nadir: the points above and below the observer.
- Meridian: the great circle that intercepts the poles, the zenith, and the nadir.

The celestial objects show an apparent motion in the sky, rising, achieving a maximum altitude above the horizon, and setting. The objects can be observable during some part of the year or could be not observable at all, at the position of the observer. Only circumpolar objects can be observed over the whole year.

The most intuitive coordinate system is the *alt-azimuth system*, based on the location of the observer. The reference plane is the local horizon; the North is the origin of the *azimuth* coordinate A that varies from 0 to 360° toward East, in analogy to the terrestrial longitude. The other coordinate, similar to the latitude, is the elevation h above the horizon, the *altitude*, that varies from 0 to 90°. The zenith distance introduced in the previous chapter is the complement to 90° of the altitude. The alt-azimuth system depends on the position of the observer and on the epoch of observation. The altitude of the star at the meridian is the maximum altitude above the horizon, the culmination. Part of the sky around the celestial pole will always be accessible to observers of one hemisphere.

The *equatorial system* is the most important coordinate system for astronomy. The equatorial coordinates of most objects are slowly varying; thus, they can be considered as fixed stars. The equatorial system is an approximation to an inertial reference system, since it is almost fixed in space. The equivalent of the longitude and latitude are the *right ascension* and the *declination*, defined by the Greek symbols α and δ. The declination measures the altitude above the celestial equator, increasing in the northern direction and decreasing in the southern direction. Right ascension starts from a reference point, the *First Point of Aries*, the point where the Sun crosses the celestial equator at the spring equinox. It is the equivalent of Greenwich longitude on Earth, increasing in the eastern direction. Right ascension and declination can be measured in angular units. The declination increases from 0 to 90° in the northern direction and decreases from 0 to −90° in the southern direction. Historically, the right ascension is measured in time units, dividing the great circle, 360°, into 24 hours. The right ascension increases from 0 to 24 h moving from the first point of Aries in the eastern direction. The meridian of the observer intercepts a different right ascension value in time. The hour angle HA is zero when the object is at meridian and changes from negative to positive values when it moves from East to West during the diurnal motion.

The equations of transformation between the two coordinate systems depend on the latitude φ of the observatory. The altitude h and the azimuth A of an astronomical source as a function of its right ascension and declination and of the hour angle are given by [3]:

$$\sin h = \sin \delta \sin \varphi + \cos \delta \cos \varphi \cos HA \qquad (2.1)$$

$$\tan A = \frac{\sin HA}{\sin \varphi \cos HA - \cos \varphi \tan \delta} \qquad (2.2)$$

The Sun appears to move along the *Ecliptic*. The Earth rotation axis is tilted by an angle of 23.4°, the obliquity of the ecliptic, with respect to the normal to the ecliptic plane. The Earth axis is slowly precessing, due to the gravitational torque of the Sun, the Moon and the planets, causing a small change in time of the equatorial coordinates. It is mandatory to state the epoch at which the astronomical coordinates are referred. The details will be discussed in the section about the definition of time in astronomy.

The *Ecliptic System* uses the ecliptic as the fundamental plane, and the first point of Aries, the intersection of the celestial equator with the Ecliptic plane, as the reference direction. The coordinates are the *ecliptic latitude* and the *ecliptic longitude*. The ecliptic latitude is the latitude above the ecliptic plane, while the ecliptic longitude is measured from the first point of Aries moving toward the East. The ecliptic system is useful for describing the motion of objects belonging to the solar system.

The *Galactic System* is different from the previous systems, since it is based on arbitrary definitions of the reference plane, the plane of the Milky Way, and of the reference direction, toward a conventionally defined galactic center. The two coordinates are the *Galactic Latitude b* and *Galactic Longitude l*. The galactic longitude increases from 0 to 360° moving toward East and starting from a reference direction pointing towards the galactic center. The Galactic Center and the Galactic North Pole are defined by the following coordinates:

- Galactic center: $\alpha = 17\,h\,42\,m\,34\,s$, $\delta = -28°55'$
- Galactic North Pole: $\alpha = 12\,h\,49\,m$, $\delta = +27°24'$

The coordinates are defined at the epoch J1950 that will be discussed below.

2.2 The Measure of Time

Astronomy has provided the first methods to measure time, using the diurnal motion of the Sun. The declination of the Sun varies in the interval between $+23°26'$ and $-23°26'$. The highest and lowest declination values correspond to the *summer solstice* and *winter solstice*, occurring approximately on June 21 and December 21. The crossings of the celestial equator correspond to the *vernal equinox* (from South to North) and to the *autumn equinox* (from North to South) that occur on March 21 and September 23; at the equinox, day and night have equal lengths. The right ascension of the Sun increases at a rate of 4 min per day. At the vernal equinox, the right ascension of the Sun is 0h, while at the autumn equinox is 12h. At the summer solstice and winter solstice, the right ascension of the Sun is 6 and 18h, respectively.

During the revolution of the Earth around the Sun, it appears to move along the ecliptic from West to East at less than one degree per day, since the Earth completes 360° in 365 days. The reference for daily timekeeping is set by the daily motion of the Sun. The *solar day* is defined as the time between two consecutive transits of the Sun at the meridian. The apparent solar time is the hour angle of the Sun. The real

Sun does not move with constant speed, because of the inclination of the ecliptic and of the ellipticity of the Earth orbit. The difference between the real and the mean Sun is described by the *Equation of Time*. The *Mean Solar Time* is the apparent time corrected for the effects mentioned above. The *Equation of Time* describes the difference between the real solar time and the mean solar time. The civil time, split into 24 h, is based on the mean solar day that assumes that the Sun is moving with constant speed across the sky.

The astronomical timekeeping [1, 2] is based in the *Universal time* (UT), the local time at Greenwich, with a time span of 24 h. The *civil time* is the mean solar time with the addition of twelve hours. The *Universal time* is the civil time at the Greenwich meridian. Local time is computed starting from Greenwich time according to the time zone of the observer.

Several definitions of the second have been adopted in time [1]. The original UT second has been originally defined as $\frac{1}{86400}$ of the mean solar day. A different definition of the second as the $\frac{1}{31556925.9747}$ of the tropical year 1900 lead to the Ephemeris Time (ET). The *Atomic Time* was defined in 1967 independently from astronomical phenomena as the time interval corresponding to 9 192 631 770 transition of ^{137}Cs. The *International Atomic Time* (TAI) was defined in 1971 as the weighted mean of different atomic clocks. The astronomical and atomic timescales were unified by the *Coordinated Universal Time* (UTC) that varies at the same pace as the TAI time, but is slightly dephased. Leap seconds are added each one or two years to synchronize the two systems. The *Terrestrial Time* (TT) is defined as the TAI time with the addition of 32.182 s. Two additional timescales are in use, referred to the geocentric barycenter and the solar system barycenter, the *Geocentric Coordinates Time* (TCG) and the *Barycentric Coordinated Time* (TCB).

When it is necessary to know the interval between two events, the Universal time is replaced by another system, the *Julian Date* (JD) system. In the Julian system, the date is computed using an origin at the First of January 4713 BC: all relevant dates for astronomy are described by positive numbers. As a convention, days are starting at noon. For reference, the Julian Day 2450000 occurred on October 9, 1995, at 12 h. To avoid long numbers, the Julian Date system is sometimes replaced by the *Modified Julian Date* (MJD) system, where: $MJD = JD - 2400000.5$.

The *epoch* is an instant in time. The astronomical standard definition of dates uses the order: year, month, day, hours, minutes, seconds, as in 2015 December 07, 00 h 39 m 00 s. The astronomical coordinates of an object, right ascension and declination, are always referred at an epoch. Several epochs have been historically used for reference:

- B1900.0, Besselian epoch J1900.000857
- B1950.0, Besselian epoch J1949.999789
- B2000, Besselian epoch J1999.998722
- B1950.000210, Besselian epoch corresponding to Julian epoch J1950.0, reference is 1950 January 1, 00 h 00 m 00 s
- B2000.001278, Julian epoch J2000, reference is 2000 January 1, 00 h 00 m 00 s

The actual reference epoch is J2000 that has superseded J1950.0.

The *Sidereal Time* (ST) is defined as the right ascension (RA) of the observer meridian and shares the units of time with the right ascension and the hour angle (HA) defined above. The three quantities are linked by the relation:

$$ST = RA + HA \qquad (2.3)$$

Due to the Earth revolution, in one year, the stars will make about 366 diurnal cycles, while the Sun will make only about 365 cycles. The sidereal time runs faster than the solar time. The length of the sidereal day is 23 h, 56 min, 4 s. Thus, the sidereal time advances two hours per month. The zero of the sidereal time can be understood recalling that the first point of Aries, the zero of the right ascension, corresponds to the vernal equinox. On that day, the sidereal time at noon is 0 h, while the solar time is 12 h. At the autumn equinox, the sidereal and the solar time at noon will be 12 h and 0 h, respectively. Since astronomical observations are performed during the night, it is necessary to deal with the sidereal time at midnight that will be 12 h at vernal equinox and 0 h at the autumn equinox. The Greenwich and local sidereal time can be obtained online using the data service at the United States Naval Observatory.[1]

2.3 Astrometry

Astrometry is a sector of astronomy dealing with the measurement of the positions of celestial sources, including the variations in time. Astrometric observations have been performed with ground-based instruments until the advent of the Hubble Space Telescope (Chap. 4). The equatorial coordinates are measured at ground-based observatories by *transit telescopes*, also called *meridian circles*. The instrument observes only the celestial objects at the meridian and is equipped with a graduated circle positioned in the circle of the meridian and a fixed reference pointing to the zenith that is used to determine the elevation angle of the pointing direction. The observations can lead to *large angle astrometry*, for the determination of the absolute positions of celestial objects, or to *small angle astrometry*, where the positions are measured relative to reference objects in the frame. The effect of the refraction and of the atmospheric turbulence (Chap. 4) limit the accuracy to some tens milli arc seconds. Optical astrometry in space is not affected by atmospheric effects. The Hipparcos mission has achieved a positional accuracy better than 1 milli arc second for the 120000 observed objects with magnitudes up to 12. The GAIA mission will perform a sky survey down to magnitude 20, with an accuracy better than 10 micro arc seconds for bright objects and better than 1 milli arc second for fainter objects.

The present reference system is the *International Celestial Reference System* (ICRS), proposed by the International Astronomical Union in 1991. The fundamental plane of the systems is the equatorial plane of the Earth, while the reference point is

---

[1] aa.usno.navy.mil/data/docs/siderealtime.php.

the vernal equinox; both of them are defined at epoch J2000.0. The ICRS is defined using extra galactic objects, the quasars, that can be considered fixed within some micro arc seconds per year. The technique of *Very Long Baseline Interferometry* (VLBI) in the radio domain (Chap. 12) allows to estimate the positions of objects with an accuracy better than 1 milli arc second. The system is a good approximation to an inertial system.

2.4 Nomenclature and Catalogs

The identification of an astronomical source and its classification rely on the historical definition of the name of stars and constellations. The telescope surveys over the whole electromagnetic spectrum are continuing adding new objects. To date, Internet resources and digital data have replaced the traditional paper and photographic storage of the first astronomical surveys. The *Centre de Donnés astronomiques de Strasbourg* (CDS[2]) is the repository dedicated to the storage, upgrading, and distribution of the information about astronomical objects. The CDS is the source of different databases:

- *Simbad*[3]: astronomical database for the identification of astronomical objects; it can be queried either by the name of the object or by the astronomical coordinates. The archive contains the main data of the sources, their identification in different surveys, and the bibliographic references.
- *VizieR*[4]: access to more than 15000 catalogs and more than 14000 data tables (May 2016) that can be queried and cross correlated. The VizieR photometry tool[5] visualizes photometric points extracted from all catalogs in VizieR to build the spectral energy distribution (SED).
- *Aladin*[6]: interactive sky atlas to access, visualize, and analyze online digitized astronomical images produced by surveys, with the possibility to superimpose data from different surveys and catalogs. Aladin is the tool used to produce the *finding charts*, the maps of the region of the sky used for observation (Chap. 13).

The Simbad utility provides access to the bibliographic references, to several measured physical quantities, and to the Aladin sky atlas. The result of the Simbad query for the quasar 3C 273 and the corresponding finding chart built with Aladin are shown in Figs. 2.1 and 2.2.

The example of Fig. 2.1 shows the rich nomenclature of astronomical names: The object has several names, since it has been observed in different surveys. The brightest stars of the constellations were classified by Bayer in 1603. They were identified with

---

[2]http://cds.u-strasbg.fr/.

[3]http://simbad.u-strasbg.fr/simbad/.

[4]http://vizier.u-strasbg.fr/index.gml.

[5]http://vizier.u-strasbg.fr/vizier/sed/.

[6]http://aladin.u-strasbg.fr/aladin.gml.

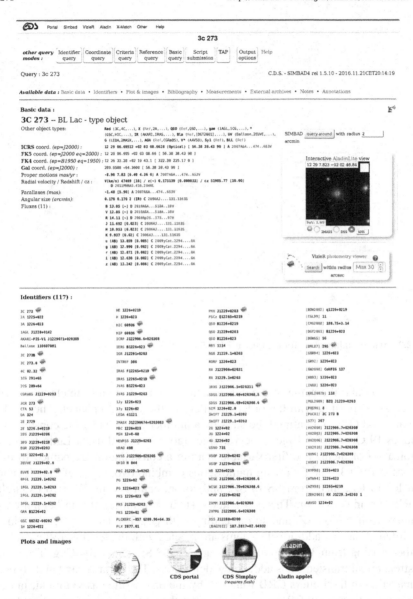

Fig. 2.1 Simbad query for the quasar 3 C 273

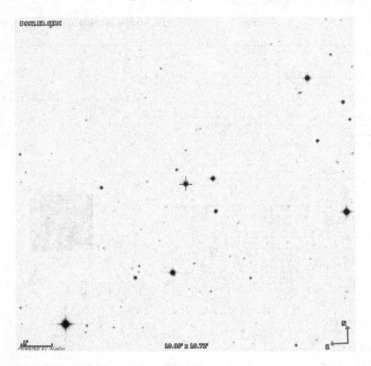

Fig. 2.2   Aladin finding chart for the quasar 3 C 273

Greek letters and the name of the constellation as in α Orionis (Betelgeuse), with brightness decreasing with the order of letters of the Greek alphabet. The catalog by Flamsteed (1725) identified the stars by a number followed by the constellation name, as 19 Orionis for Rigel. The variable stars are named using Latin letters or combination of them. The first variable stars in a constellation are named from R to Z, as in R CrB. The following name series combine two letters of the alphabet, from RR to RZ, then from SS to SZ, and so on, ending with ZZ; two examples are RR Lyr and SS Cyg. The two letter system then moves to the beginning of alphabet, with AA to AZ, and so on, ending with QZ. The letter system produces 334 different names; additional systems are labeled with the letter V followed by a number starting from 335; two examples are V5558 Sgr and V404 Cyg. The world of astronomical transients has added new definitions. The gamma-ray bursts (GRB) are named with the Letters GRB followed by the date written as yymmdd; in case there is more than one GRB per day, those after the first one get an additional letter. Supernovae are named using the letters SN followed by the year written as yyyy and by a capital letter, from A to Z and, if needed, starting again from aa. The recent discovery of gravitational waves announced in 2016 has added a new definition for a new astronomy, GW for gravitational waves, followed by the date of the event in the format yymmdd. The first reported event is GW150914.

The first star catalog, realized by Bessel during the nineteenth century, included a few tens of stars. During the twentieth century, it was replaced by the *Fundamental Katalog* (versions FK4 and FK5) with about 1500 stars. More recent catalogs have introduced new classifications, which is made of a set of letters and a progressive number. Some of them will be briefly summarized here. The symbols BD, CD label the *Bonner Durchmusterung* and *Cordoba Durchmusterung*, two photographic surveys of the two hemispheres up to magnitude 10. The label HD labels the *Henry Draper* Catalog, originated by a donation to the Harvard Observatory, that includes more than three hundreds thousands stars up to magnitude 9. The Schmidt telescope at Palomar has been used for the *Palomar Observatory Sky Survey* (POSS) of 1957, a photographic survey including stars from the North Celestial Pole down to declination -33° and up to magnitude 20; the survey was completed by the Australian survey that extended the declination range to $-90°$. The label USNO refers to the catalog of the *United States Naval Observatory* that has digitized the POSS and other surveys and includes about one billion stars up to magnitude 21. Additional catalogs are built by ground-based and space-based observatories, by adding new prefixes to the nomenclature.

2.5 Internet Resources

The amount of Internet resources for astronomers is steadily evolving, and any attempt to build an exhaustive compilation would become out of date very soon. The present section is a road map addressing the general topics of data collection and storage and the development of management tools that have lead to the birth of the Virtual Observatory. The data sets produced by a large number of sky surveys and space missions are available online and contain the potential for new discoveries. The Simbad, Aladin, and VizieR facilities discussed above are high-level data products that are interfaced to a rich world of archives. Another high-level product is the NASA/IPAC Extragalactic Database (NED)[7] that combines the imaging and spectral data of millions of objects external to the Milky Way.

Some widely used astronomical archives are:

- Palomar Observatory Sky Surveys (POSS I and POSS II): the photographic plates which have been digitized and are available at The STScI Digitized Sky Survey site[8]
- Sloan Digital Sky Surveys (SDSS-I,[9] SDSS-II,[10] SDSS-III[11]): a series of massive spectroscopic surveys providing images and spectra for millions of sources.

---

[7]https://ned.ipac.caltech.edu/.
[8]http://stdatu.stsci.edu/cgi-bin/dss_form.
[9]http://www.sdss.org/.
[10]http://www.sdss2.org/.
[11]http://sdss3.org/.

- NASA/IPAC Science Archive[12]: archive collecting the data of the infrared surveys and missions (2MASS, WISE, Herschel, Planck, IRAS).
- Mikulski Archive for Space Telescopes (MAST)[13]: access to the astronomical data archives of optical, ultraviolet, and near-infrared observatories, including the Hubble Space Telescope (HST) and the Digitized Sky Survey (DSS).
- High Energy Astrophysics Science Archive Research Center (HEASARC)[14]: repository of X-ray and gamma ray missions (RXTE, Swift, Rosat, Chandra, XMM-Newton, Fermi, INTEGRAL, etc.).

The data of archives generally are proprietary for a time interval of one or two years, before becoming available to the whole community. The *Virtual Observatory* (VO)[15] is the environment managing the electronic access to the archives of ground- and space-based observatories. The effort has lead to the development of different software tools addressed to query the world archives that can run as stand-alone programs or as Web-based utilities:

- VOPlot[16]: tool for the analysis and the graphing of data.
- Tool for OPerations on Catalogues And Tables (TOPCAT)[17]: an interactive graphical environment to plot and analyze data
- SPLAT-VO[18]: tool for display and analysis of astronomical spectra.
- Specview[19]: for display and loading of spectra and comparison with libraries of spectra.
- VOSpec[20]: spectral analysis tool with libraries of atomic and molecular data.

The astronomical literature can be directly accessed at the SAO/NASA Astrophysics Data System (ADS)[21], a portal supporting three bibliographic databases. For each paper, ADS links to the publisher site for the full text or to a scanned versions; the systems also tracks the citations to the papers. The electronic archive *arXiv*[22] has replaced the old paper preprints and provides the prepublication version, freely downloadable, of most astronomical papers of the last two decades.

---

[12]http://irsa.ipac.caltech.edu/frontpage/.

[13]https://archive.stsci.edu/.

[14]http://heasarc.gsfc.nasa.gov/.

[15]http://www.ivoa.net, http://www.euro-vo.org/.

[16]http://vo.iucaa.ernet.in/~voi/voplot.htm.

[17]http://www.star.bris.ac.uk/~mbt/topcat/.

[18]http://www.g-vo.org/pmwiki/About/SPLAT.

[19]http://www.stsci.edu/institute/software_hardware/specview/.

[20]http://www.cosmos.esa.int/web/esdc/vospec.

[21]http://www.adsabs.harvard.edu/.

[22]http://arxiv.org.

Problems

2.1 Using Simbad, Vizier, and Aladin find the astronomical coordinates and build the finding charts for the following objects, discussing the type of source and the main properties:

- HD 209458
- Barnard's star
- SS Cyg
- M67
- OJ 287
- Crab

References

1. Lèna, P. et al.: Observational Astrophysics. Springer-Verlag Berlin Heidelberg (2012)
2. Smarr, W. H. M.: Textbook on Spherical Astronomy. Cambridge University Press (1977)
3. Bely, P.: The Design and Construction of Large Optical Telescopes. Springer, New York (2013)

Part II
Optical Astronomy

Chapter 3
Optical Astronomy: Telescopes

The optical telescope is the symbol of astronomy, the collector of radiation from astrophysical sources. The telescopes and their aberrations will be firstly discussed using the formalism of geometric optics. Present telescopes use reflecting optical elements, generally two mirrors, in different configurations to reduce the aberrations. The improvement of the light collection capability is proportional to the aperture; thus, larger and larger telescopes are being designed and built, with a steady advancement in the manufacturing techniques of large mirrors.

A telescope collects the radiation from the astronomical source through an aperture. The *entrance pupil* is the limiting aperture encountered by the incoming radiation, whose size determines the brightness of the final image. The *exit pupil* is the image of the entrance pupil produced by the optical components behind it. Optical elements, such as lenses and mirrors, are characterized by the focal length f. The ratio of the focal length to the diameter D of the aperture is the *focal ratio* $F = \frac{f}{D}$. Systems with small ($f/2$) and large ($f/8$) focal rations are called *fast* and *slow* optical systems. Since the image size scales as the focal length, while the collection capability scales as the square of the aperture, the image brightness scales as the reciprocal of the square of the focal ratio. Thus, fast telescopes produce a brighter image, but a smaller magnification. Telescope is characterized by two additional quantities. The *image scale* is related to the number of resolving elements on the detector (pixels for CCDs, millimeters for photographic plates). The *field of view* is the angular region on the sky observable with the telescope.

Detailed discussions of optical telescopes can be found [1–3, 5–7].

3.1 Telescope Mounts

The *telescope mounts* are used to point the telescope in the direction of a celestial object and to track its motion in the sky. The mount types (Fig. 3.1) are based on the

© Springer International Publishing Switzerland 2017
R. Poggiani, *Optical, Infrared and Radio Astronomy*,
UNITEXT for Physics, DOI 10.1007/978-3-319-44732-2_3

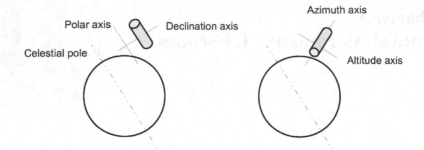

Fig. 3.1 Working principle of equatorial and alt-azimuth mounts

alt-azimuth and equatorial coordinate systems described in Chap. 2. The alt-azimuth mount that controls the elevation above the horizon and the azimuth is technically simpler, since the telescope weight has a constant orientation with respect to the axes bearings. Tracking the star motion requires the driving of both vertical and horizontal axes at the same time with variable speed. The equatorial mount ensures a fixed orientation with respect to the sky. One axis, the *polar axis*, is pointed to the North Celestial Pole, parallel to the Earth axis and rotated at the rate of one revolution per sidereal day to counteract the Earth rotation. The other axis, the *declination axis*, is orthogonal to the polar axis.

The historical large telescopes mostly used equatorial mounts. The *German mount* has been widely used. The axis of declination is mounted on the polar axis, that is fixed to a pillar. The telescope is balanced by a counterweight. The advent of computer technology has introduced the use of alt-azimuth mounting for large telescopes, getting advantage of their easier mechanical design.

3.2 Refracting Telescopes

The refracting telescope, the first telescope in history, uses two optical elements, the *objective* (a lens) and the *eyepiece* (one or more lenses). The refracting telescopes how some disadvantages and have been replaced by the reflecting telescopes. The optical lenses show *chromatic aberration*, since the refraction index of optical materials depends on the wavelength λ (Fig. 3.2). The focal length of a lens is a function of the wavelength. At each point on the optical axis, there will be a clear image at one wavelength, surrounded by blurred images at different wavelengths. The chromatic aberration can be reduced by combining optical refracting elements with different power, as the *achromatic doublet*, the coupling of a biconvex crown glass lens with plano-concave flint glass lens that brings radiation of two different wavelengths at the same focus. The chromatic aberration is proportional to the area of the aperture; thus, it precludes the realization of large refracting telescopes. The largest refracting telescope is the Yerkes telescope, with an aperture of 40 in.

Fig. 3.2 Refraction index of different optical materials, data available at http://refractionindex.info/

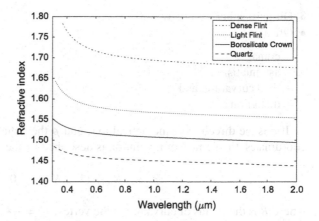

3.3   Geometrical Optics: Optical Aberrations

The large telescopes in operation are based on reflecting elements, one or more mirrors. There is no chromatic aberration, since the reflection does not depend on the wavelength. The reflectivity of different layer materials in the optical and near-infrared spectral regions is reported in Fig. 3.3. Silver and aluminum coatings are suitable choices for the optical region.

The reflecting telescope is made of conical surfaces produced by conic section rotating about the revolution axis [1–3, 5–7]. The normal to the conic in each point bisects the angle formed by the rays that connect the point with the two foci; thus, the rays that arise in a focus converge to the other focus. In principle, a point-like source is imaged into a perfect image (stigmatism). However, the presence of *aberrations* produces a blur. The natural choice for a conic is the parabola: all rays from the focus at infinity and parallel to the optical axis are brought to the close focus of the parabola. The optical aberrations of a conical surface are [5]:

Fig. 3.3 Reflectance of aluminum, silver, and gold, data from [4]

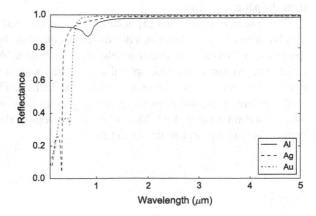

- On-axis: spherical aberration.
- Off-axis

  - coma,
  - astigmatism,
  - field curvature, and
  - distortion.

If z is the direction of the optical axis and ρ the radial coordinate in cylindrical coordinates, the surface of revolution is described by the conic vertex equation:

$$\rho^2 - 2Rz + (1 + K)z^2 = 0 \tag{3.1}$$

where R is the radius of curvature at the vertex, $K = -e^2$ is the conic constant, and e is the eccentricity of the conic. The different surfaces are defined by the value of the conic constant: K is null for a sphere, is equal to -1 for a paraboloid, is smaller than -1 for an hyperboloid, is positive for a prolate ellipsoid, and is in the range between -1 and 0 for an oblate ellipsoid. The vertex equation can be expanded in a Taylor series:

$$z = \frac{\rho^2}{2R} + (1 + K)\frac{\rho^4}{8R^3} + \cdots \tag{3.2}$$

In the following, third-order aberrations will be considered. The focal length of a conical mirror for a ray incident parallel to the optical axis at a distance ρ from the axis is [5]:

$$f(\rho) = \frac{R}{2} - \frac{1 + K}{4R}\rho^2 - \frac{(1 + K)(3 + K)}{16R^3}\rho^4 + \cdots . \tag{3.3}$$

The focus of paraxial rays is at a distance $\frac{R}{2}$ from the vertex. The rays incident at a distance ρ from axis are directed to the same focus only in a paraboloid mirror, where $K = -1$. The rays for spherical or ellipsoidal surfaces are focused closer to the vertex of the conics, while for hyperboloid surfaces the rays are focused farther than the paraxial focus.

The *spherical aberration* [5] is caused by the different foci of paraxial and marginal (external) rays and includes two different contributions, the *transverse spherical aberration* (TSA) and the *angular spherical aberration* (ASA). The first effect is related to the distance from the optical axis at the location of the paraxial focus for a ray incident at a radius ρ and reflected off the surface. The second effect is related to the difference between the reflection angles of rays incident on a generic conic surface and on a paraboloid. The two contributions at the third order are inversely proportional to a power of the focal ratio:

$$TSA = -(1+K)\frac{\rho^3}{2R^2} \tag{3.4}$$

$$ASA = -(1+K)\frac{\rho^3}{R^3} \tag{3.5}$$

The spherical aberration can be eliminated using a paraboloid mirror, for which $K = -1$. The most famous spherical aberration has been found in the Hubble Space Telescope mirror; the defect was later corrected with the installation of the COSTAR system by a dedicated Shuttle mission.

The other aberrations listed above are off-axis aberrations related to rays incident with an angle θ with respect to the optical axis. The cylindrical coordinate system is replaced by a system with standard Cartesian coordinates, with ρ replaced by y. The total contribution of off-axis aberrations (AA) at the third order is a sum of terms proportional to $\theta^m y^n$, with $n + m = 3$ [5]:

$$AA = 3a_1\frac{y^2\theta}{R^2} + 2a_2\frac{y\theta^2}{R} + a_3\theta^3 \tag{3.6}$$

The first term is the *coma*. Incident rays cross different sections of the focal plane; thus, the final image has an asymmetrical comet shape, with a tail pointing outward; the degradation increases with increasing distance from the axis. The effect of the coma is inversely proportional to the square of focal ratio; thus, it is stronger in fast telescopes. The second term is the *astigmatism*: the focus of the rays in the plane of optical axis is focused at different distances, forming foci with a segment shape. Astigmatism is inversely proportional to the focal ratio. The third term is the *distortion*: equidistant points on a grid appear shifted with respect to each other, in a barrel or pincushion-like deformation. Finally, there is the effect of *field curvature*: rays with different inclination angles are focused on curved surfaces.

The aberrations in reflecting telescopes set limits on the observable field. The main telescope systems are based on one mirror on a combination of two mirrors. In two-mirror systems, there is an additional contribution to aberrations from the misalignment between the primary and secondary mirror.

3.4 Single-Mirror Telescopes

The simplest reflecting telescope uses one mirror only, shaped as a paraboloid. The system is called a *prime focus* system. The instrumentation is mounted at the paraboloid focus and produces a partial obscuration of the aperture. The 5-m Palomar telescope is a prime focus system. The spherical aberration of prime focus telescopes is zero, unlike coma and astigmatism. The residual aberrations are a function of the focal ratio F [5]:

$$coma = \frac{\theta}{16F^2} \tag{3.7}$$

$$astigmatism = \frac{\theta^2}{2F} \tag{3.8}$$

Coma is dominant at small angles, settings the limits to the field of view. The prime focus telescope has a large optical efficiency since they use a single reflecting surface.

3.5 Two-Mirror Telescopes

According to the *Schwartzschild Theorem*, it is possible to correct for n aberrations using n optical elements [1, 5]. The compensation of aberrations by the combination of two or more mirrors relies on the shift of the wave front induced by each surface that can in principle compensate the effect of the previous one. A single mirror can correct only for a single aberration. Two mirrors can be used to correct for spherical aberration and coma, in principle. The two mirrors are the *primary*, that determines the aperture, and the *secondary*. The focus of the second conical surface is coincident with the focus of the first conical surface. The basic two-mirror telescopes are the *classical telescopes*, shown in Fig. 3.4. The *Cassegrain* configuration uses a paraboloid primary mirror and a convex hyperboloid secondary mirror; the exit pupil is a virtual image behind the secondary mirror. The *Gregorian* configuration uses a paraboloid mirror and an ellipsoidal secondary mirror; the exit pupil is an image between the primary and the secondary. The Cassegrain configuration is more compact.

The properties of the two-mirror telescopes will be discussed using the Cassegrain telescope as an example. The Cassegrain system is characterized by the following physical quantities [5]: y_1 and y_2 are the radii of primary and secondary mirror; R_1 and R_2 are the curvature radii at the vertex of primary and secondary; s_2 and s'_2 are the object and image distance of the intermediate object and of the final image as measured from the vertex of the secondary mirror; and f_1 is the focal length of the primary. The system can be described by defining a set of normalized parameters:

- $k = \frac{y_2}{y_1}$: the ratio of the mirror radii that is also the ratio of the marginal ray heights;

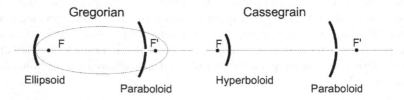

Fig. 3.4 The Gregorian and Cassegrain telescopes

- $\rho = \frac{R_2}{R_1}$: the ratio of the vertex radii;
- $m = -\frac{s_2'}{s_2}$: transverse magnification of the secondary; and
- $f_1\beta$: the back focal distance, the distance of the focus from the vertex of the primary mirror.

By applying the reflection equations to the surface of the secondary mirror, it is possible to show that:

$$m = \frac{\rho}{\rho - k} \tag{3.9}$$

$$k = \frac{\rho(m - 1)}{m} \tag{3.10}$$

$$\rho = \frac{mk}{m - 1} \tag{3.11}$$

$$1 + \beta = k(m + 1) \tag{3.12}$$

The focal ratio of a two mirror Cassegrain telescope is:

$$F = mF_1 \tag{3.13}$$

where F_1 is the focal ratio of the primary mirror. The two-mirror configuration allows to build compact systems, by folding large focal lengths.

The combination of two mirrors allows the correction of the spherical aberration. One example is the *classical Cassegrain* telescope, where the primary mirror is a paraboloid and the secondary mirror is an hyperboloid, with conic constants:

$$K_1 = -1 \tag{3.14}$$

$$K_2 = -\left(\frac{m + 1}{m - 1}\right)^2 \tag{3.15}$$

The design of a Cassegrain telescope requires the definition of the parameters m, the magnification of the secondary, and β that determines the distance of the telescope focus from the vertex of the primary mirror. The values of m, β determine the values of ρ, the ratio of mirror radii of curvature, and of k, the ratio of mirror radii. The final step is the choice of the radius y_1 of the primary mirror, the effective aperture of the two-mirror telescope, and of its focal length f_1, driven by observational constraints.

The residual aberrations of the classical telescopes are [5]:

$$coma = \frac{\theta}{16F^2} \tag{3.16}$$

$$astigmatism = \frac{\theta^2}{2F} \frac{m^2 + \beta}{m(1 + \beta)} \tag{3.17}$$

The coma is identical to that of the single-mirror paraboloid telescope, while the astigmatism is higher by a factor of about m. Coma is the main contribution limiting the field of view.

The simultaneous modification of the conic constants K_1 and K_2 of the primary and secondary mirrors, according to [5]:

$$K_1 + 1 = \frac{k^4}{\rho^3} \left[K_2 + \left(\frac{m+1}{m-1} \right)^2 \right] \tag{3.18}$$

defines the family of Cassegrain telescopes with null spherical aberration. The telescopes where both coma and spherical aberration are absent are called *aplanatic telescopes*. The *Ritchey–Chretien* telescope is an aplanatic Cassegrain telescope with hyperboloid primary and secondary mirrors [1, 5]:

$$K_1 = -1 - \frac{2(1+\beta)}{m^2(m-\beta)} \tag{3.19}$$

$$K_2 = -\left(\frac{m+1}{m-1} \right)^2 - \frac{2m(m+1)}{(m-\beta)(m-1)^3} \tag{3.20}$$

The Ritchey–Chretien is the standard configuration for large telescopes and has been adopted for the Hubble Space Telescope. The residual aberration is the astigmatism [1, 5]:

$$astigmatism = \frac{\theta^2}{2F} \left[\frac{m(2m+1)+\beta}{2m(1+\beta)} \right] \tag{3.21}$$

The field is limited by the astigmatism, but is larger than the field of classical telescopes.

An additional contribution to the aberrations is given by the errors in the alignment of the secondary mirror with respect to the primary. The secondary mirror can be tilted and/or located off-center with respect to the optical axis of the primary, introducing coma and astigmatism and shifting the image orthogonally to the optical axis. The contributions at the third order of these aberrations are inversely proportional to a power of the focal ratio.

Telescopes are equipped with different foci configurations [1]. Usually, large telescopes have several foci with different instruments that are permanently mounted, to operate the desired one without any intervention. Single-mirror telescopes operate as prime focus instruments or with a Newtonian focus, with a small inclined mirror to direct radiation outside the instrument tube. The *prime focus* telescopes usually have fast focal ratios, but have a limited field of view and set strong constraints on the instrument dimensions. The *Cassegrain focus* is a very popular solution. The access is easy, instruments can be large and massive, and the optical losses are reduced since only two optical surfaces are involved. The *Nasmyth focus* is an external focus of the Cassegrain type fixed to the telescope tube, realized using a mirror inclined by 45°. It is used for leaving instruments in site when the telescope is using other foci. The *Coudè focus* is a variation where an additional mirror sends the radiation to a location separate from the telescope; the largest and heavier instruments, such as spectrographs, are often mounted at Coudè foci.

The focal ratio at the telescope foci is driven, among other factors, by the cost of the telescope assembly that is related to the length of the telescope tube [1]. The primary mirror should have a fast focal ratio, a solution that has progressively become more popular in modern telescopes. Fast primary mirrors have also the advantage of being smaller and less massive. The limiting factor of fast systems is given by the misalignment between the primary and the secondary mirror.

3.6 Three-Mirror Telescopes and Beyond

The addition of mirrors can in principle produce the cancelation of three aberrations [1]. Some configurations are summarized here for completeness. The *Paul-Baker telescope* has a paraboloid primary mirror, and spherical secondary and tertiary mirrors with identical curvature radii to cancel the spherical aberration. The systems show also null coma and null astigmatism. The absence of the third-order aberrations produces high-quality images, but the configuration offers a limited space for the installation of instrumentation, in particular spectrographs. Four-mirror telescopes give an even larger freedom in reducing the aberrations. The solution is of interest for large telescopes with a spherical primary mirror that can be machined and polished more easily. The four-mirror telescope is the afocal combination of a spherical primary with a paraboloid secondary mirror. Coma and astigmatism are eliminated, unlike spherical aberration. The addition of a third spherical mirror and of a fourth mirror cancels the spherical aberration, at the price of partially reintroducing coma and astigmatism. Detailed analysis of the variety of multiple mirror telescope configurations and of the related aberrations can be found in the books listed in the bibliography.

3.7 Advanced Telescope Systems

The mirrors for large telescopes [1, 3, 6, 7] are built as monolithic mirrors or with segmented elements. The ideal material for the blank of a large mirror must be dimensionally stable, easily machinable, with low density, weakly sensitive to thermal deformations and with a fast thermal constant. The mass of the mirror must be minimized to reduce the deformation under gravity in ground-based facilities and to reduce the launch costs for space-based instruments. The deformation under self-weight is proportional to $\frac{\rho}{E}\frac{D^4}{t^2}$, where E is the Young modulus, ρ is the density, and D and t are the diameter and thickness of the mirror [1]. The properties of some blank mirror materials [1] are reported in Table 3.1. The mechanical performances

Table 3.1 Physical properties of mirror blank materials, based on [1]

Material	ρ (kg/m^3)	E (GPa)	α (10^{-6}/K)	C (J/kg K)	k (W/m K)
Aluminum	2700	70	23	890	170
Borosilicate glass	2200	63	3.3	800	1.2
ULE-fused silica	2200	68	0.03	760	1.3
Zerodur	2500	91	0.05	820	1.5
Silicon carbide	3200	466	23	730	190

of the mirror are determined by the specific thickness $\frac{E}{\rho}$. The thermal properties are governed by the quantity $\frac{k}{C\alpha\rho}$, where k is the thermal conductivity, C is the heat capacity, and α the thermal expansion coefficient. The blank material must have the lowest possible thermal coefficient to reduce the sensitivity to temperature variations. Borosilicate glass has been the traditional choice in the past, due to its low thermal expansion coefficient. The *Ultra Low Expansion* (ULE) fused silica and Zerodur have replaced borosilicate glass, thanks to their lower thermal expansion coefficient. Silicon carbide offers the advantages of high stiffness, but is fragile. Aluminum shows a high thermal conductivity, but also a high coefficient of thermal expansion; the main advantage is the low cost.

The large telescopes rely on the mechanical design of the mirrors and of the mounting systems to reduce the deformations of the optical surfaces that occur during the pointing and the tracking of the sources. Since the most relevant source of wavefront error is the misalignment of the secondary mirror, the holders of the primary and secondary mirrors are mounted on an element that is the central part of the telescope tube, the *Serrurier truss*, in an open arrangement to reduce the effect of the wind and to facilitate the thermalization of the primary mirror with the environment. The truss maintains the alignment of the mirrors independently from the pointing direction of the telescope [1].

The largest monolithic mirror has a diameter of 6 m and a weight of about 42 tons and is mounted at the BTA-6 telescope.[1] The previous largest mirror has been the 5-m mirror of the Hale telescope at Palomar,[2] that has been made lighter by the presence of tens cavities in the bulk. The cost of a monolithic mirror is proportional to a power of the diameter. The blank materials discussed above are expensive, and machining becomes more difficult with the increasing of the mirror size. In addition, the deflection due to the weight is proportional to the fourth power of the diameter. The deformations induced by gravity are reduced using *honeycomb mirror* that combines a low weight with a high stiffness. Two sheets joined by a hollow structure, usually made of hexagonal cells, are produced by molding. The low mass

---

[1]http://www.sao.ru.

[2]http://www.astro.caltech.edu/palomar/observer/P200/observers.html.

Fig. 3.5 Layout of a segmented mirror

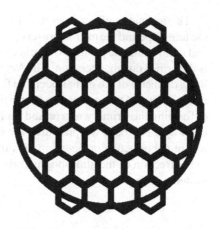

allows more compact and less expensive mount and driving systems than those of monolithic mirrors. The high stiffness allows to maintain the shape of the optical surface against the mechanical and thermal deformations. An additional advantage of the reduction in the mass of the mirror is the decrease of thermal inertia: the mirror achieves thermal equilibrium more quickly than an equivalent monolithic mirror. The honeycomb strategy has been adopted for the Hubble Space Telescope.

A large mirror can be realized by the combination of separate mirrors, as in the *Multimirror Telescope* (MMT),[3] that consisted of six telescopes with apertures of 1.8 m and separate secondaries, sharing a common mount. The instrument had an equivalent aperture of 4.5 m and has been operative from 1979 to 1998. The multimirror combination has strong demands on the accuracy of the relative positions of its elements and requires a beam combiner to send the light to a common focus.

The mirrors of the largest telescopes are realized by assembling different elementary components or segments (Fig. 3.5). The first *segmented mirror* made of 61 elements and with an aperture of 1.5 m was developed by Horn D'Arturo in 1952.

The segments, generally hexagonal in shape, are assembled to build mirrors with conical surfaces. The 10-m mirror of the *Keck telescope*[4] has been the first large segmented mirror, made of 36 hexagonal elements with a side of 0.9 m. Each segment is equipped with actuators and edge sensors. The signals of the sensors are used to drive the actuators that correct the effects of gravity and thermal deformation, with a frequency of tens Hz. The Keck telescope is in operation since 1993 and has been the inspiration for other segmented and actively controlled instruments that will be discussed in Chap. 4.

---

[3]http://www.mmto.org.

[4]http://www.keckobservatoty.org.

The technique of *active optics* corrects the misalignment and the deformation of the telescope and the mirror, preserving the shape of the primary mirror at all positions and controlling the position of the secondary mirror. The shape of a *meniscus mirrors* is deformed during operation. The *New Technology Telescope* (NTT),[5] whose mirror has a diameter of 3.6 m, has been the first instrument with active optics. A wavefront sensor (Chap. 4) measures the properties of the wavefront and corrects its error by deforming the primary mirror and tuning the position of the secondary mirror. The technique has been adopted also for the *Very Large Telescope* (VLT) array.[6] The technique of active optics maintains the figure of the mirror, but does not perform the correction of the atmospheric turbulence, that is achieved by the adaptive optics systems (Chap. 4).

Problems

3.1 Compare the equatorial and the alt-azimuth mounts.

3.2 Discuss the effects of the different aberrations on the image produced by a reflecting telescope.

3.3 A two-mirror telescope has a primary with a diameter of 4.2 m and a focal ration f/2 and a secondary with a diameter of 0.6 m. Estimate the obscuration factor and the minimum magnification in the Cassegrain configuration to have the focus behind the primary mirror. After choosing a value of the magnification ratio, estimate the distance of the focus from the primary mirror, the ratio of the curvature radii of the mirrors, and the curvature radius of the secondary.

3.4 Discuss the motivations for using fast focal ratios in large telescopes.

References

1. Bely, P.: The Design and Construction of Large Optical Telescopes, Springer, New York (2003)
2. Oswalt, T. D., Bond, H. E.: Planets, Stars and Stellar Systems. Volume II: Astronomical Techniques, Software, and Data. Springer (2013)
3. Oswalt, T. D., McLean, I. S.: Planets, Stars and Stellar Systems. Volume I: Telescopes and Instrumentation. Springer (2013)
4. Paquin, R. A.: Properties of Metals. In: Bass, M., Van Stryland E. W., Williams D. R., Wolfe W. L (eds.) Handbook of Optics, McGraw-Hill (1995)
5. Schroeder, D.: Astronomical Optics. Academic Press (1999)
6. Wilson, R. N.: Reflecting Telescope Optics II. Springer (2002)
7. Wilson, R. N.: Reflecting Telescope Optics I. Springer (2007)

---

[5]http://www.eso.org/facilities/lasilla/telescopes/ntt.html.
[6]http://www.eso.org/paranal.

Chapter 4
Telescopes: Ground Based or in Space?

This chapter discusses the optical telescopes including the diffraction linked to the wave nature of the electromagnetic radiation. The diffraction determines the response function of a telescope and sets the limit to the angular resolution that can be achieved. Ground-based telescopes are affected by the atmospheric turbulence and do not achieve the diffraction limit, that is, achieved by space-based telescopes. The effects of the atmosphere can be partially compensated by the technique of adaptive optics, by deforming the telescope mirror to match the wavefront error. The chapter is closed by the discussion of large telescope facilities.

4.1 Physical Optics: Diffraction Theory

The previous chapter has discussed optical telescopes in the context of geometrical optics, where the image of a point-like object is point-like, unless aberrations are present. Due to the wave nature of light, the image of a point has a finite extension. The *diffraction* of light is based on the Huygens–Fresnel Principle. Each point of a wave front is the source of secondary spherical waves that interfere with each other, producing a new wave front. The image of a star formed by a telescope is described by the *Fraunhofer diffraction*, since the source is at large distance from the diffracting aperture. The formalism of diffraction is discussed in detail in [5] and will be briefly summarized here. Considering a wave front emerging from an aperture (Fig. 4.1), the amplitude $U(P)$ of the diffracted wave at the point P with coordinates (x, y, z) on the screen is [5]:

$$U(P) = constant \int A(\eta, \xi) \exp[-ik(p\xi + q\eta)]d\xi d\eta \qquad (4.1)$$

where (ξ, η, ζ) are the coordinates of a point on the aperture; p, q are the director cosines $\frac{x}{R}, \frac{y}{R}$; $A(\xi, \eta)$ is the aperture function that describes the transparent and opaque regions of the aperture. The integral is a two-dimensional Fourier integral,

© Springer International Publishing Switzerland 2017
R. Poggiani, *Optical, Infrared and Radio Astronomy*,
UNITEXT for Physics, DOI 10.1007/978-3-319-44732-2_4

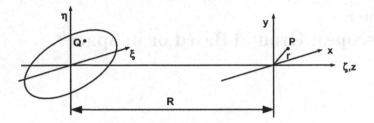

Fig. 4.1 Coordinate system for the aperture and the screen in the Fraunhofer diffraction

thus, the amplitude of the diffracted wave is the Fourier transform of the aperture function. The measured quantity is the intensity, $I = |U(P)|^2$.

Two families of apertures are of special interest for astronomical instrumentation. The first one is the rectangular aperture with sides $2a$ and $2b$ that produces an intensity of the diffracted wave given by the product of two sinc functions [5]:

$$I = I_0 \left(\frac{\sin kpa}{kpa} \right)^2 \left(\frac{\sin kqb}{kqb} \right)^2 \tag{4.2}$$

where k is the wave vector. The function $sinc^2(x) = (\frac{\sin x}{x})^2$ is shown in Fig. 4.2, left. The intensity has a principal maximum at $x = 0$ and zero minima at $x = \pm m\pi$. The first minima are at $p = \frac{\lambda}{2a}, q = \frac{\lambda}{2b}$. The full width at half maximum along one direction is of the ratio of the wavelength to the width of the aperture along the same direction.

The second relevant example for astronomy is the circular aperture with a diameter $D = 2a$, the standard aperture of telescopes. The intensity of the diffracted wave is given by:

$$I = I_0 \left[\frac{2J_1(kar/R)}{kar/R} \right]^2 \tag{4.3}$$

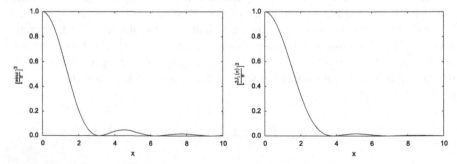

Fig. 4.2 *Left*: diffraction from a rectangular aperture, plot of $\left(\frac{\sin x}{x} \right)^2$; *right*: diffraction from a circular aperture, plot of $\left(\frac{2J_1(x)}{x} \right)^2$

where R is the distance of the screen from the diffraction pattern and r the radial distance of a point on the diffraction pattern. The function $(\frac{2J_1(x)}{x})^2$ shows a bright central maximum at $x = 0$ surrounded by null minima and secondary maxima (Fig. 4.2, right).

4.2 The Point Spread Function

The *Point Spread Function* (PSF) is a response function defined as the intensity distribution of the diffraction pattern produced by a point-like source normalized to its maximum value [1, 5]. The fundamental equation of image formation in telescopes is a convolution equation:

$$O(l, m) = \int \int P(l - l', m - m')B(l', m')dl' \, dm' \qquad (4.4)$$

where B is the intrinsic source brightness distribution, O the observed brightness distribution, P is the Point Spread Function, and l, m are the coordinates on the sky. According to the convolution theorem, the Fourier transform of the observed brightness distribution $\hat{O}$ is the product of the Fourier Transforms of the Point Spread Function $\hat{P}$ and of the intrinsic brightness distribution $\hat{B}$:

$$\hat{O}(u, v) = \hat{P}(u, v) \times \hat{B}(u, v) \qquad (4.5)$$

where u, v are spatial frequencies, related to the variation of intensity across the image. The *Optical Transfer Function* (OTF) is the Fourier Transform of the Point Spread Function and describes the properties of the optical system and the spatial frequency components of the final image. The real part of the optical transfer function is called *Modulation Transfer Function* (MTF). All contributions to the response of an optical system can be modeled by a suitable optical transfer function. The total transfer function of a ground-based observing system is the product of the transfer function of the telescope and of the transfer function of the atmosphere. For circular symmetry and no aberrations, the transfer function will be a function of the combination of spatial frequencies, $v = \sqrt{u^2 + v^2}$. The modulation transfer function tends to the unit for vanishing spatial frequencies and to 0 for frequencies close to the cutoff frequency of the system. The extension of the approach from point sources to extended sources is straightforward, since they can be decomposed into a set of individual point sources. The concept of Point Spread Function will be extensively used in the following.

4.3 The Airy Pattern and Diffraction Limited Telescopes

The Point Spread Function of real optical systems cannot be determined analytically, but is estimated using the information on the wavefront errors at the aperture. The Point Spread Function of a perfect optical system free of aberrations and modeled as a circular aperture is the *Airy pattern*, a function of the angular distance θ of a point on the diffraction pattern. Since the source and the screen are at large distances, the angle is small and approximately equal to the ratio of the radial coordinate r on the pattern to the screen distance R. The Airy pattern (Fig. 4.3) is thus given by [1–5]:

$$I = \left[\frac{2J_1(v)}{v}\right]^2 \tag{4.6}$$

where $v = \frac{\pi D\theta}{\lambda}$, D is the aperture diameter. The first minimum occurs at the angle $\theta = 1.22\frac{\lambda}{D}$: The radius of the first dark ring defines the size of the *Airy disk*, of the order of $\frac{\lambda}{D}$.

The more general case of interest for astronomy is a partially obscured aperture, as in a telescope with a primary mirror with radius a and a secondary mirror with a radius $b = \varepsilon a$. The Point Spread Function for a partially obscured aperture is [5]:

$$I = \frac{1}{(1-\varepsilon^2)^2}\left[\frac{2J_1(v)}{v} - \varepsilon^2\frac{2J_1(\varepsilon v)}{\varepsilon v}\right]^2 \tag{4.7}$$

The profile of the Point Spread Functions of two apertures, one unobscured ($\varepsilon = 0$) and one with partial obscuration ($\varepsilon = 0.35$), is reported in Fig. 4.4.

The Point Spread Function concept provides a criterion to assess the quality of an image, through the *encircled energy* (EE) [1, 5], the energy enclosed within a defined radius v_0; for a circular aperture, the encircled energy is given by:

Fig. 4.3 The Airy pattern, the Point Spread Function of a perfect optical system with a circular aperture

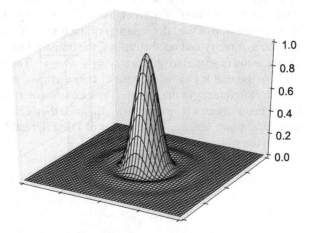

Fig. 4.4 Point Spread Function of circular apertures: unobstructed aperture (*solid line*) and obstructed aperture with an obscuration ratio of 0.35 (*dashed*)

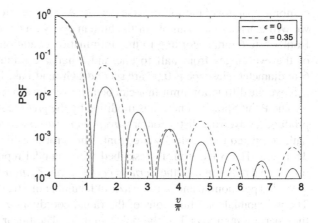

Fig. 4.5 Encircled energy of a circular aperture

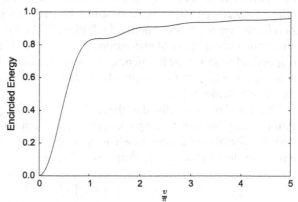

$$EE_u(v_0) = 1 - J_0^2(v_0) - J_1^2(v_0) \qquad (4.8)$$

The first, second, and third dark rings contain the 83.8, 91, and 93.8 % of the total energy, respectively (Fig. 4.5). The partial obscuration of the aperture shifts some energy from the Airy disk to the first ring.

The Point Spread Function is the starting point to define the *resolving power* of an optical system, the ability to discriminate the images of distinct point-like objects, and the angular resolution of the observation. The *Rayleigh criterion* states that in systems limited by diffraction, the images of two point sources with identical intensity will be resolved if the principal maximum of the first image is coincident with the first minimum of the second image. The angular separation corresponding to the Rayleigh criterion is:

$$\theta = 1.22\frac{\lambda}{D} \qquad (4.9)$$

where D is the diameter of the aperture. An alternative criterion for angular resolution is the *Sparrow criterion*, where the limit is given by the lack of minimum in the combined pattern of the sources and $\theta = \frac{\lambda}{D}$. As an example of the Rayleigh criterion, we will consider the example of the human eye, with a pupil diameter that changes from 1.5 to 6 mm according to the illumination conditions. The angular resolution of the eye ranges from half to one and a half arc minute. The resolution of the 6-m diameter telescope is 0.02 arc seconds. Ground-based telescopes are not able to achieve the diffraction limit that can be achieved by space-based telescopes.

The Point Spread Function is modified by the presence of aberrations, since they produce a wavefront distortion or *wavefront error* related to the difference of optical path compared to the ideal wave front. The error is estimated using the rms of the deviation. The wave front is described with a modal representation, using a set of orthogonal polynomials, the *Zernike polynomials $Z_j(n, m)$*, where n is the degree of the radial polynomial and m is the azimuth number of a trigonometric function [1, 2]. The polynomials are functions of the radial coordinate r, the radius normalized to the aperture radius, and of the polar angle ϕ. The linear superposition of polynomials $\Sigma a_n Z_n(r, \phi)$ allows to reconstruct the error of a generic wave front. The first Zernike polynomials are reported in Table 4.1. The low-order Zernike polynomials correspond to the classical aberrations discussed in Chap. 3. Higher order terms correspond to higher spatial frequencies. A combination of the first few tens polynomials is generally sufficient to describe the surface, deformation, and alignment errors of the optical elements [1].

The *Strehl ratio* is defined as the ratio of the on-axis peak value of the instrument Point Spread Function to the peak of the ideal diffraction limited Point Spread Function. The Strehl ratio decreases when the system shows aberrations. If the aberrations have a residual variance $(\Delta\phi)^2$, the Strehl ratio is:

$$S = \exp[-(\Delta\phi)^2] \tag{4.10}$$

Table 4.1 The first Zernike polynomials

Symbol	Name	Polynomial
Z_1	Piston	1
Z_2	x-Tilt	$2r \cos \phi$
Z_3	y-Tilt	$2r \sin \phi$
Z_4	Focus	$\sqrt{3}(2r^2 - 1)$
Z_5	0° Astigmatism	$\sqrt{6}r^2 \cos 2\phi$
Z_6	45° Astigmatism	$\sqrt{6}r^2 \sin 2\phi$
Z_7	x-Coma	$\sqrt{8}(3r^3 - 2r) \cos \phi$
Z_8	y-Coma	$\sqrt{8}(3r^3 - 2r) \sin \phi$
Z_9	x-Trifoil	$\sqrt{8}r^3 \cos 3\phi$
Z_{10}	y-Trifoil	$\sqrt{8}r^3 \sin 3\phi$
Z_{11}	Third-order spherical	$\sqrt{5}(6r^4 - 6r^2 + 1)$

The ratio is equal to 1 for a diffraction limited image. An optical system is considered *diffraction limited* if the value of the Strehl ratio is $S \geq 0.8$, corresponding to an error on the wave front of $\frac{\lambda}{14}$ (Marechal criterion).

In view of discussing the effect of the atmosphere on the observations, the properties of the telescope as an optical system will be summarized using the modulation transfer function. The transfer function of an unobscured circular aperture with a diameter D is given by [5]:

$$T(\nu) = \frac{2}{\pi} \left[\cos^{-1} \nu_n - \nu_n \left(1 - \nu_n^2 \right)^{\frac{1}{2}} \right] \qquad (4.11)$$

where ν is the spatial frequency, ν_n is the normalized spatial frequency:

$$\nu_n = \frac{\nu}{\nu_c} \qquad (4.12)$$

where ν_c is the cutoff. In the diffraction limit, $\nu_c = \frac{D}{\lambda}$, where D is the telescope diameter. The MTF of a circular aperture is reported in Fig. 4.6. The presence of the obscuration produces a drop of the MTF in the intermediate frequency range and an increase close to the cutoff frequency, since the corresponding PSF is narrower and has a larger amount of energy in the first ring compared to the Airy pattern. The effect of optical aberrations is estimated by including the suitable aperture function in the Fourier integral [5].

The optical transfer function governing a ground-based astronomical observation is the product of the telescope transfer function hereby defined and of the atmospheric transfer function that will be discussed in the next section.

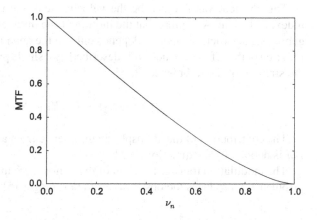

Fig. 4.6 Modulation transfer function of a circular aperture

4.4 The Effect of the Atmosphere: Seeing

Ground-based telescopes do not achieve the diffraction limit, due to the atmospheric turbulence that mixes air regions with different temperature and produces fluctuations of the refractive index [1–5]. The wave front of a star is plane parallel at the entrance of the atmosphere, but is deformed before arriving at the telescope aperture, with a major impact on the formation of the image. The main effects occurring at a light collection aperture are:

- Seeing: random variations of the direction of the radiation entering the aperture that produces a broadening of a star image;
- Scintillation: random intensity fluctuations that cause the twinkling of the star light[1]

The turbulent process starts at a large scale L (*outer scale*, tens to hundreds meters), transferring energy to a cascade of smaller and smaller eddies, down to the smallest scale l (*inner scale*, of the order of millimeters), where viscous dissipation occurs. The *inertial range*, the region between the inner and the outer scale, is modeled by the *Kolmogorov theory*. The Kolmogorov spectrum of kinetic energy is a function of the wave vector k. It behaves as $k^{-\frac{5}{3}}$ for isotropic turbulence in one dimension and as $k^{-\frac{11}{3}}$ for isotropic turbulence in three dimensions. The spatial structure of the turbulence is described by the structure function of a suitable physical quantity a linked to the random changes, i.e., to the expectation value of the difference of the values of the quantity measured at two different points R_1, R_2.

$$D_a(R_1, R_2) = < |a(R_1) - a(R_2)|^2 > \qquad (4.13)$$

The physical variable can be the velocity field, the temperature, the refraction index, and so on. Assuming that the motion of the turbulence is isotropic and homogeneous, the structure function depends only on the quantity $|R_1 - R_2|$. The structure function of the refraction index n is described by a single parameter, C_n^2, summarizing the strength of the turbulence [2, 5]:

$$D_n(R_1, R_2) = C_n^2 |R_1 - R_2|^{\frac{2}{3}} \qquad (4.14)$$

The contribution of the atmospheric turbulence from a given layer with thickness Δh is defined by the quantity $C_n^2 \Delta h$.

The modulation transfer function of the atmosphere in the Kolmogorov model of turbulence reproduces the observed profile of images [5]:

$$T_t = exp\left[-3.44 \left(v_n \frac{D}{r_0}\right)^{\frac{5}{3}}\right] \qquad (4.15)$$

---

[1] Source of inspiration for many poets.

The transfer function is very close to a Gaussian and produces a quasi Gaussian Point Spread Function, with a Gaussian core and wings with a less steep decay. A more suitable PSF that describes the effects of the atmosphere is the Moffat function:

$$I = \frac{I_0}{[1 + \left(\frac{\theta}{R}\right)^2]^\beta}, \tag{4.16}$$

The quantity r_0 is the *Fried parameter* that depends on the turbulence strength C_n^2 and provides a quantitative estimation of the atmospheric quality:

$$r_0 = 0.185\lambda^{\frac{6}{5}}(\cos z)^{\frac{3}{5}}\left[\int C_n^2(h)dh\right]^{-\frac{3}{5}}, \tag{4.17}$$

where z is the zenith distance. Due to the dependence on the wavelength, the Fried parameter is larger in the infrared. The value of r_0 or, equivalently, of C_n^2, is the driver for the choice of the observatory site. The integral estimates the integrated effect of the turbulence. Assuming a value of $\int C_n^2(h)dh \sim 10^{-12}$ cm$^{\frac{1}{3}}$ for a typical observatory site and a wavelength of at 500 nm, the value of r_0 is of the order of 0.1 m.

A snapshot of the turbulence secured at high speed will show a pattern of speckles, with an angular size of the order the diffraction limit $\frac{\lambda}{D}$. The speckles move inside a circular region with an angular diameter of about $\frac{\lambda}{r_0}$. The number of speckles is of the order of $(\frac{D}{r_0})^2$. The atmospheric turbulence produces a broadening of star images, called *seeing*. An astronomical observation with a long exposure will measure an integrated effect, i. e., the wandering of the speckle pattern over the envelope region. The *seeing disk* is defined by the width of the Point Spread Function that includes the effect of the atmosphere and is of the order of $\frac{\lambda}{r_0}$. Thus, the size of the star image changes from the value $\sim\frac{\lambda}{D}$ in the diffraction limit to $\sim\frac{\lambda}{r_0}$ in the seeing limit. The angular size of a star image does not depend on the value of the aperture, for apertures with diameters larger than the Fried radius r_0. Thus, the Fried parameter sets the resolution of ground-based telescopes, that is the same resolution of a diffraction limited telescope with aperture r_0. Observations secured with an aperture smaller than the Fried radius probe a single diffraction limited pattern moving at the timescale of the turbulence. The angular resolution corresponding to the Fried radius is of the order of one arc second, much worse than the diffraction limit. For example, a 10-m telescope shares the resolution of a small telescope with an aperture of about 10 cm.

The atmospheric quality of ground-based sites is evaluated using image quality as a benchmark, with a continuous monitoring of atmospheric turbulence over long time intervals. Large facilities are equipped with *seeing monitors* [1]. A class of seeing monitors uses small telescopes with a diameter of some tens of centimeters equipped with a detector to measure the sizes of the star images. The family of the *Differential Image Monitors* measures the differential motion of the image of a star observed through two different atmospheric paths and directly probes the Fried radius.

The image quality is affected also by the intensity variation across the aperture, produced by rays diffracted by the turbulence that undergo interference. The scintillation is produced by the deformation of the wave front that causes a fluctuation of the detected intensity. The effect is the cause of the twinkling of the star light observed by the naked eye. The scintillation affects observing systems with an aperture smaller than the Fried radius, that include also the human eye. Telescopes with increasing larger apertures average the effect.

4.5 Adaptive Optics Systems

The technique of *adaptive optics* corrects the wavefront distortion due to the atmosphere by the introduction of an opposite distortion [2, 5]. An adaptive optics system includes the following blocks (Fig. 4.7):

- a *deformable mirror* that produces the conjugate of the incoming distorted wave front;
- a *wavefront sensor* that measures with a high sampling rate distortion of the wave front observing a close star, natural or artificial; and
- a *control system* that uses the information of the wavefront sensor to produce a deformation of the mirror for correcting the atmospheric effects.

The adaptive optics systems are constrained by the timescale of the turbulence and by the extension of the region where reference stars can be chosen. The atmospheric turbulence can be modeled by regions with constant phase and dimension r_0. According to the Taylor model of *frozen turbulence*, each turbulence region is considered frozen while crossing the telescope aperture transported by the wind. The timescale of turbulence is related to the Fried radius r_0 and to the wind speed v_w:

$$\tau_0 = \frac{r_0}{v_w} \tag{4.18}$$

For $r_0 \sim 0.1$ m, $v_w \sim 10$ m/s, the timescale τ_0 is of the order of 10 ms. The velocity of the adaptive optics corrections is of the order of some hundreds Hz. The radiation from two stars separated by an angular distance θ passes through different regions and undergoes a different phase change. If the turbulent layer is at an altitude h, the *isoplanatic angle* defines the maximum angle of separation that allows the light from the two stars to experience the same phase variation:

$$\theta_0 \sim \frac{r_0}{h} \tag{4.19}$$

Assuming a value of the Fried parameter of the order of 0.1 m and an altitude of a few kilometers for the turbulent layers, the isoplanatic angle is of the order of a few arc seconds. The reference star must be very close to the object of interest.

The requirements on the adaptive optics controls are governed by the ratio of the telescope aperture to the Fried radius, $\frac{D}{r_0}$. The formalism of the Zernike polynomials is useful to estimate the residual aberration after correcting a fixed number of modes

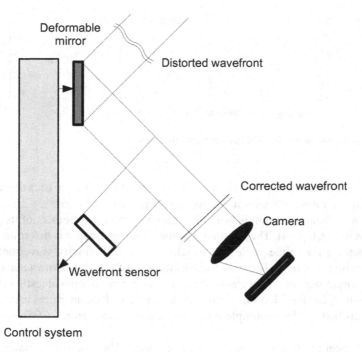

Fig. 4.7 Layout of an adaptive optics system

with adaptive optics. The dominant contribution is produced by the first terms, starting from the tip–tilt. The residual phase variance that is left after the correction of the first j modes in a circular aperture is, for large j [2, 5]:

$$\Delta_j = 0.2944\, j^{-\frac{\sqrt{3}}{2}} \left(\frac{D}{r_0} \right)^{\frac{5}{3}} \tag{4.20}$$

After the adaptive optics correction, the Point Spread Function shows a diffraction limited core with a width $\sim \frac{\lambda}{D}$ emerging on top of the wider seeing limited profile, with a width $\sim \frac{\lambda}{r_0}$.

The adaptive optics corrections require a guide star, natural or artificial [1–5]. The sky coverage is limited by the availability of reference stars that must be within the isoplanatic patch of the target. The guide stars must be within the isoplanatic angle, of the order of a few microradians at most. The *natural guide stars* provide only a partial sky coverage at visible wavelengths, with some improvement in the infrared region. The artificial *laser guide stars*, realized with the scattering of laser light in the high layers of the atmosphere, provide the access to the whole sky. The Rayleigh laser beacons use the Rayleigh scattering occurring at an altitude of 10–20 km, above the layers with the stronger contribution of turbulence. The scattering on the neutral sodium layer occurs at a height of about 90 km is a better approximation of a star at

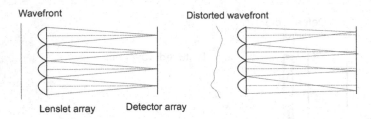

Fig. 4.8 Principle of the Shack–Hartmann wavefront sensor

infinity. The *Multiconjugate Adaptive Optics* (MCAO) technique uses several guide stars and several wavefront sensors, together with different mirrors.

The wavefront distortion is generally sensed by the measurement of its gradient, the wavefront tilt [1–4]. The sensitivity of the wavefront sensor is determined by the photometric error on the guide star: to achieve a high sensitivity, wavefront sensors operate with white light. The most used instrument is the *Shack–Hartmann wavefront sensor*, a modification of the classical Hartmann system for optical testing (Fig. 4.8). The sensor splits the telescope aperture into a matrix of subapertures using an array of lenslets that produce multiple images on a suitable detector, a CCD, in the focal plane.

The lenslet array is positioned at telescope pupil. The size of the lenslets is of the order of the Fried radius. The size of the subapertures is of the order of a few times the seeing value to ensure that the images are always contained. The subapertures defined by the lenslets, typically 2×2, produce star spots on the CCD, whose centroids are shifted by the tilt fluctuations of the wave front. The shifts allow to estimate the wave front distortion. The centroid of the intensity distribution with respect to the direction i depend on the partial derivative of the wave front $\frac{\partial \phi}{\partial i}$, on the focal length f, on the wavelength and on the subaperture size l:

$$C_i = \frac{\lambda f}{2\pi} \int_0^{\frac{l}{2}} \int_0^{2\pi} \frac{\partial \phi}{\partial i} \rho d\rho d\theta \qquad (4.21)$$

Since the detection principle is purely geometrical, the operation of a Shack–Hartmann wavefront sensor does not depend on the wavelength of the radiation and can use broadband radiation to achieve a better sensitivity.

Since the largest contribution to the wavefront distortion is due to the tilt modes, they are corrected separately from the higher order Zernike ones. The tip and tilt of the wave front can be compensated by a *tip–tilt mirror* that deals with the motion of the image caused by the atmospheric turbulence, the vibrations of the telescope, the effect of the wind, and so on.

The deformable mirror (Fig. 4.9) corrects the wavefront distortion measured by the wavefront sensor by adjusting its own shape by the application of suitable force or momenta [1–4]. The mirrors of adaptive optics systems consists of different elements coupled to actuators. There are two main families of deformable mirrors, the *contin-*

Fig. 4.9 Layout of a mirror for adaptive optics systems

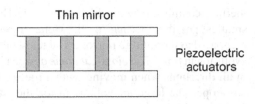

Thin mirror

Piezoelectric actuators

uous mirrors and the *bimorph mirrors*. Continuous mirrors are thin plates equipped with arrays of piezoelectric or, more rarely, electrostrictive actuators below the surface. The actuators provide the pushing and pulling of the surface. The continuous mirrors provide an intrinsically continuous surface, but the actuators must operate in parallel and not independently. The solution is very effective to correct higher order modes, while the tilt is corrected by a tip-tilt mirror. Bimorph mirrors are sandwiches of piezoelectric materials (PMN, PZT ceramic) equipped with electrodes. The application of a voltage difference between the layers produces a contraction in one layer and an expansion in the other one, thus a bending of the assembly. The bimorph mirrors are fabricated more easily than the continuous ones; thus, they are less expensive.

We close this section by underlining that the adaptive optics systems partially compensate for the effects of the atmosphere, but they are dependent on the site seeing at the observatory site through the Fried radius. The technique can be successfully implemented at observatory sites with high sky quality.

4.6 Technical Issues for Ground- and Space-Based Telescopes

The performance of astronomical observations at a telescope is driven by the signal-to-noise ratio that will be discussed in detail in Chap. 13. Intuitively, the signal is proportional to the collection area, while the noise is proportional to the dimension of the image of the object. Thus, it is natural to build larger telescopes, but at the same time, it is necessary to reduce the image size. Ground-based telescopes can achieve larger apertures and can be maintained more easily than space-based telescopes. The performances are limited by the atmospheric turbulence that degrades the Point Spread Function. The use of adaptive optics provides a partial compensation of the atmospheric effects. Space-based telescopes operate at the diffraction limit. In addition, the sky background is smaller.

Telescopes operating on Earth are installed inside an enclosure [1–4], but well above the ground level, since the atmospheric surface layers have a strong effect on seeing. The thickness of the surface layer is of the order of a few meters; thus, the floor of the telescope is places at an height of 10–20 m at least. The *dome seeing* is produced by convective air motions, if the air inside the dome is not in thermal equilibrium with the external air. The telescope mirrors should have the smallest possible thermal inertia. Honeycomb mirrors are intrinsically advantaged compared to monolithic mirrors. The telescope enclosures have undergone a steady evolution in time. The *classical dome* design is large enough to contain the telescope in all directions when moving, with a roof that can be rotated independently of the telescope. The large air volume inside the classical dome requires techniques for flushing, with large roof slits equipped with a shutter that is tuned according to the wind conditions. The control room is located below the telescope level. The larger telescopes such as MMT use the *corotating dome* solution. The control and the service rooms are built on the sides of the telescope assembly, reducing the size of the enclosure. While the first large telescopes were operating with a typical seeing of about two arc seconds, the removal of heat sources from the dome has allowed to approach the limit of one arc second. The modern thermal and mechanical design techniques have given seeing levels better than 0.5 arc seconds.

The space-based telescopes have the major advantage of operating at the diffraction limit [1–4]. The mirrors must have the lowest possible mass, due to the cost of the space access, but also a great strength to go through the launch sequence and maintain the alignment. For example, the primary mirror of the HST is made of an honeycomb core fused between two plates. The telescope on board is accompanied by a set of several dedicated instruments, such as photometers and spectrographs, that are the topics of the next chapters. The space telescopes demand a large number of ancillary systems for operation, among them the systems for pointing and for thermal control. The pointing and control of the telescope and related instrumentation is critical, due to the small size of the Point Spread Function. The pointing is performed using star tracking techniques with cameras that compare the position of the stars in the observed field with stars in reference catalogs. The orientation of the spacecraft is sensed through a set of gyroscopes. Electronic systems are dedicated to the communication with ground, for the management of controls and for the transmission of data. The stability of operation of a space telescope is ensured by the thermal control system that manages the protection of the payload from the solar heat and the stability of the operation temperature of instruments. Active thermal controls include combinations of coolers, usually with cryogens, and heaters, with a mission lifetime of the mission limited by the lifetime of the cryogens. Passive thermal controls rely on the insulation of components with shields, complemented to arrays of radiating panels.

4.7 Ground- and Space-Based Facilities

The choice of the Earth site of an astronomical observatory requires a detailed investigation of the local seeing. The criterion of low seeing holds not only for standard optical telescopes, but also for system equipped with adaptive optics. The astronomical observatories are built at sites with a good seeing, smaller than 1 arc sec, at altitudes that are above the temperature inversion layer. The turbulence at an observatory site is given by the contribution of different atmospheric layers. The turbulence in the *surface layer* is governed by the wind shear on local irregularities of the terrain and by the presence of close trees. The thickness of this layer ranges from a few meters at good observatory sites to some tens meters at an average site: the telescope should be built above the surface layer. The *planetary boundary layer* includes the part of the atmosphere where the air motion due to convection occurs and extends up to the *inversion layer*, at an altitude of about 1 km. Clouds are stopped by the inversion layer: large observing facilities are at high altitude, above the inversion layer. The *atmospheric boundary layer* is above the planetary boundary layer. The convection is blocked by the inversion layer, but modulations in the region surface can trigger the formation of gravity waves. The observatory should be located close to the sea or the ocean exposed to the dominant winds. The highest layer is the *free atmosphere*, where there is no residual effect from the ground irregularities. The turbulence in this layer is concentrated close to the jet streams at high altitude. Islands and coastline sites are both suitable choices for observatory sites. They should be preferably at tropical latitudes, since the cloud coverage is small and the velocity of winds in the upper part of the atmosphere is lower. The island sites should be very far from the mainland and, possibly, with a single high mountain emerging above the inversion layer. The mountains in islands offer a stable atmospheric environment and have been chosen for the observatories at Mauna Kea (Hawaii) and at the Canary Islands. Sites on the coast benefit from the low altitude of the inversion layer due to the low temperature of the sea. If there is a mountain with the top above the inversion layer and dominant winds coming from the sea or ocean direction, the same conditions of islands sites are reproduced. This is the environment of observatories in Chile and in California. Antarctica is a singularity for the global wind circulation; thus, the upper atmosphere is very stable, from the point of view of the turbulence; in addition, it is also thermally stable.

The choice of orbits for space-based telescopes is driven by the compromise between the science and the cost. The extremely successful Hubble Space Telescope (HST) is in a low orbit at an altitude of some hundreds kilometer. The forthcoming James Webb Space Telescope will be operating at the L_2 Lagrangian point, as WMAP and GAIA.

4.8 Large Telescopes

The necessity of large telescopes is explained by the dependence of the signal-to-noise ratio of the observations on the source flux and on the aperture. The details will be discussed in Chap. 13. The number of photons collected by a telescope is proportional to the source flux density and to the aperture, proportional to the mirror diameter. The ideal telescope should have the largest possible aperture and operate as close as possible to the diffraction limit. In the following, some examples of large telescope facilities in operation or being designed are presented. The instrumentation installed at the telescopes will be discussed in the next chapters.

The 10-m mirror of the *Keck telescope* at Observatory Hawaii discussed in Chap. 3, has been the first large telescope with a segmented mirror, made of 36 hexagonal segments with a side of 0.9 m that form an hyperboloid surface. Keck telescope has been joined by a second identical instrument, *Keck-2*, at the same site. The two instruments can operate as an interferometer with a baseline of 75 m. The *Gran Telescopio Canarias* (GTC)[2] that started the operation in 2007 shares the design of Keck telescopes with its 36 segments and an aperture of 10 m.

Honeycomb primary mirrors have been adopted for several facilities. The *Sloan Digital Sky Survey* (SDSS)[3] telescope uses an honeycomb primary mirror with a diameter of 2.4 m. The second-generation *Multiple Mirror Telescope*[4] uses a 6.5 m honeycomb mirror and two secondaries, one standard with a 1.7 m diameter and another belonging to the adaptive class. The twin *Magellan Telescopes*[5] at Las Campanas, Chile, use 6.5-m honeycomb mirrors as primaries and adaptive optics. The *Large Binocular Telescope* (LBT)[6] uses two honeycomb mirrors with a diameter of 8.4 m installed on the same mount to achieve an equivalent aperture of 11.8 m. It is the combination of two Gregorian telescopes pointing at the same region and equipped with active and adaptive control systems. The two telescopes can be used separately, but their light can also be combined for phased array imaging of the field.

The *Hobby–Eberly telescope* (HET)[7] at the McDonald Observatory is made of 91 spherical mirror segments with 1.5 m size, with a total aperture of 9 m. The instrument is fixed in elevation and can move in azimuth only. The direction of the mirror does not change with respect to gravity, releasing the demands on the mount system and cutting the overall cost of the system. The telescope is used as a prime focus instrument. The *South African Large Telescope* (SALT)[8] is a similar instrument on the Southern hemisphere that started operation in 2005. The *Large Sky Area Multiobject Fiber Spectroscopic Telescope* (LAMOST)[9] has a spherical primary mirror with 6-m aperture consisting of 37 segments with 1.1 m size.

The best example of a mission operating in space is the *Hubble Space Telescope* (HST),[10] launched in 1990 in a low Earth orbit and still in operation as of 2016. The primary mirror has a diameter of 2.4 m. The different instruments on board provide a coverage of the spectrum from 0.12 to 24 μm, extending beyond the optical region into the ultraviolet and the infrared. The multipurpose approach has made HST

---

[2]http://www.gtc.iac.es/.

[3]http://www.sdss.org/.

[4]https://www.mmto.org/.

[5]http://obs.carnegiescience.edu/Magellan.

[6]http://www.lbto.org/.

[7]http://www.as.utexas.edu/mcdonald/het/het.html.

[8]https://www.salt.ac.za/.

[9]http://www.lamost.org/public/?locale=en.

[10]https://www.nasa.gov/mission_pages/hubble/main/index.html.

an extremely successful observatory, that is continuing to produce a rich wealth of results. The low orbit has allowed servicing with the Space Shuttle.

The *Kepler* mission[11] has been launched in 2009 in an Earth trailing and heliocentric orbit. Kepler has a telescope with an aperture of 0.95 m equipped with a wide field ($10^0 \times 10^0$) CCD camera. The mission is devoted to monitor the stars in the field with high photometric precision to search for extrasolar planets by the transit method. To date, thousands of candidates have been found.

The forthcoming ground-based telescopes will combine large apertures with adaptive optics techniques. The *Large-Scale Synoptic Survey* (LSST)[12] will use a 8.4-m-diameter mirror. Facilities with apertures in the range from 20 to 40 m are being designed. The *Giant Magellan Telescope* (GMT),[13] with an effective aperture of 24.5 m, will be built at Las Campanas in Chile. The GMT will be an aplanatic Gregorian and will be modeled on the principle of the MMT, with seven circular mirrors with a diameter of 8.4 m. It will have a secondary mirror, a concave mirror with a diameter of 3.2 m with seven segments matching the corresponding elements of the primary mirror. An adaptive mirror with deformable elements or a fast-steering mirror with rigid but mobile segments will allow both diffraction limited and seeing limited observations. The *Thirty Meter Telescope* (TMT)[14] will be a Ritchey–Chretien telescope, but without a Cassegrain focus. A tertiary mirror will send the radiation to the instrumentation at two Nasmyth foci. As the GMT, the TMT will allow both diffraction and seeing limited observations, but the secondary mirror is not equipped with active or adaptive controls. The primary mirror will consist of 472 segments with 0.715 m size with sensors and actuators as in the Keck telescopes. The *European Extremely Large Telescope* (E-ELT)[15] will be an anastigmatic instrument with three mirrors and two additional mirror for folding. The primary mirror will have a diameter of 39.42 m and will be built assembling segments with a size of 1.45 m. The secondary and tertiary mirrors will have diameters of 6 and 4.2 m, respectively.

The forthcoming space-based telescopes include the *James Webb Space Telescope* (JWST),[16] the heir of the Hubble Space Telescope, scheduled for launch toward the L2 point in 2018. The instrument will be launched in a folded arrangement (with a dimension of 4.5 m) and will be deployed in orbit. The primary mirror will have an aperture of 6.5 m and will composed of 18 segments, made of beryllium that is lighter than the standard optical materials. The operation in the space environment, with the absence of gravity and the reduced thermal effects, does not require an active control of the optical surface. The mirror will not be equipped with sensors and actuators. The *Advanced Technology Large Aperture Telescope* (ATLAST) is a forthcoming instrument with a segmented primary mirror with a diameter of 16.8 m that aims to observe in the region from 0.1 to 2.5 μm.

---

[11] http://kepler.nasa.gov/.

[12] http://www.lsst.org/.

[13] http://www.gmto.org/.

[14] http://www.tmt.org/.

[15] https://www.eso.org/sci/facilities/eelt/.

[16] http://www.jwst.nasa.gov/.

Robotic telescopes play a special role in optical astronomy, since they are used for two different tasks: the systematic survey of the sky and the follow-up of transients detected by optical instruments, but also by high energy instruments. The monitoring of the sky allows to identify a large number of known and new sources, building meaningful statistical samples of different classes of objects and classifying their properties. The detection of transients in the domain of high energy photons or the newly discovered gravitational wave events is followed by the search and the characterization of the optical counterpart. The robotic telescopes usually have relatively small apertures, less than 2 m, but a large field of view to cover the sky in the shortest possible time. Instruments dedicated to the follow-up of transients require fast-slewing mounts, to be able to start the observation within seconds from the dissemination of the event. The *All Sky Automated Survey* (ASAS)[17] uses two wide field CCDs a 25 cm telescope, imaging the whole sky is imaged within a few days. The *SuperWASP* instrument[18] is an array of eight wide field systems that can image the whole sky in less than one hour. The original *Robotic Optical Transient Search* (ROTSE)[19] project used four wide field imagers mounted on a fast-slewing mount for the follow-up of transients that lead to the first identification of an optical counterpart of a gamma-ray burst. The *ROTSE-III* instrument is a network of fast-slewing telescopes with 45-cm aperture distributed over four continents and able to provide a 24-h monitoring of the sky. The *RAPid Telescopes for Optical Response* (RAPTOR) is a global network of fast response telescopes targeted to the prompt follow-up of transients, in particular gamma-ray bursts, with different instruments for monitoring the sky with high cadence (20 s) and for prompt response. The *Pi of the Sky*[20] monitors a large part of the sky with a high cadence. The *Mobile Astronomical System of Telescope Robots* (MASTER)[21] is network of 40-cm telescopes aiming to survey the sky down to magnitude 20. The *Rapid Eye Mount* (REM) and the *Telescopio Ottimizzato per la Ricerca dei Transienti Ottici Rapidi* (TORTORA) are fast-slewing instruments for the study of the early afterglows of gamma-ray bursts. The *Optical Gravitational Lensing Experiment* (OGLE)[22] search for microlensing has been performed with a 1.3-m telescope with a field of view of 1.5° equipped with a mosaic of CCDs. The *Liverpol Telescope* at La Palma is a robotic telescope with an aperture of 2 m that has a time allocation policy open to a wide community. The telescope has an aperture of 2 m and is equipped with an optical imager, a spectrometer, a polarimeter, and infrared instrumentation. The *Sloan Digital Sky Survey* (SDSS) has used the 2.5-m telescope at Apache Point, securing multiband photometry of more than two thousands millions objects and more than one million spectra. All data are available online.[23] The imaging instrumentation is made of 30 CCDs

---

[17] http://www.astrouw.edu.pl/asas/.

[18] http://www.superwasp.org/.

[19] http://rotse.net/.

[20] http://www.pi.fuw.edu.pl/.

[21] http://observ.pereplet.ru/.

[22] http://ogle.astrouw.edu.pl/.

[23] http://www.sdss.org/; http://www.sdss2.org/; http://sdss3.org/.

with filters designed for the survey (see discussion in Chap. 6). The *Palomar Transient Factory* (PTF)[24] at the Oschin Telescope (72 in aperture) uses a mosaic of six CCDs and monitors the sky in different modes, among them the search for transients (cadence of a few days) and the dense sampling of selected events (with a cadence of minutes); transients are detected by comparison with template images, classified with Bayesian methods and possibly followed-up at other telescope facilities. The *Panoramic Survey Telescopes And Rapid Response System* (PanSTARRS)[25] project is a set of four telescopes with 1.8-m aperture (with a field of view of 3°) observing the same region of the sky at the same time, equipped with a mosaic of 60 CCDs. The instruments described so far have contributed to the detection of the afterglows of the gamma-ray bursts, to the detection of supernovae and other transients, and to the detection and the classification of thousands variable stars.

Problems

4.1 Discuss the main technical aspects involved in the building of large telescopes.

4.2 Discuss the effects of the atmosphere on ground-based observations.

4.3 Estimate the Fried parameter at an observing facility with a 10-m telescope and a seeing of 0.5 arc seconds measured at the wavelengths of the visible radiation (500 nm). Estimate the coherence time assuming a wind speed of 10 m s^{-1}. Estimate the value of the Fried parameter in the near infrared at 2.2 μm.

4.4 Discuss the differences in adaptive optics systems using natural and laser guide stars.

4.5 Compare the active optics and the adaptive optics strategies.

4.6 Estimate the angular resolution of the Hubble Space Telescope that has a primary mirror with a diameter of 2.4 m and is operating at the diffraction limit, in the visible region (500 nm).

References

1. Bely, P.: The Design and Construction of Large Optical Telescopes, Springer, New York (2003)
2. Lèna, P. et al.: Observational Astrophysics. Springer-Verlag Berlin Heidelberg (2012)
3. Oswalt, T. D., McLean, I. S.: Planets, Stars and Stellar Systems. Volume I: Telescopes and Instrumentation. Springer (2013)
4. Oswalt, T. D., Bond, H. E.: Planets, Stars and Stellar Systems. Volume II: Astronomical Techniques, Software, and Data. Springer (2013)
5. Schroeder, D.: Astronomical Optics. Academic Press (1999)

---

[24]http://www.ptf.caltech.edu/.

[25]http://pan-starrs.ifa.hawaii.edu/public/.

Chapter 5
Optical Astronomy: Detectors

This chapter discusses the instruments used to detect the radiation collected by the optical telescopes and transform it into an electrical signal. The main classes of detectors for optical radiation will be presented. All detectors have a fundamental limit caused by the statistical properties of the electromagnetic radiation. Historically, optical astronomy has initially used photographic plates, with imaging capability, but a nonlinear response. The photomultiplier tubes have a linear response, but are single pixel detectors. The charge-coupled devices (CCDs) have revolutionized the field, with their linear response and the possibility to perform multiobject photometry in a single frame. To date, the CCDs are the most used detectors at telescopes and will be discussed in detail. The next chapters will show their applications in astronomical photometry and spectroscopy.

5.1 Detectors: The Basics

Astronomical detectors for radiation in the optical, infrared, and radio regions can be divided into three families:

1. *Photon detectors*: The incident photons release electrons in the interaction with the detector material. Photon detectors are used over a large part of the electromagnetic spectrum, from the gamma rays to the infrared region.
2. *Thermal detectors*: The photon energy is converted into heat and changes some physical property of the detector. The process is intrinsically broadband and has found a wide application in the infrared and submillimeter detectors.
3. *Wave detectors*: Measurement of the electric field of the incident electromagnetic wave is, generally used in radio astronomy

© Springer International Publishing Switzerland 2017
R. Poggiani, *Optical, Infrared and Radio Astronomy*,
UNITEXT for Physics, DOI 10.1007/978-3-319-44732-2_5

The present chapter discusses the photon detectors used in optical astronomy. Thermal detectors are discussed in the context of infrared astronomy (Chap. 8) and submillimeter astronomy (Chap. 9). Wave detectors will be presented in Chap. 9. While photon and thermal detectors are incoherent, since they can only measure the signal amplitude, wave detectors are coherent, since can measure the polarization and the phase of the wave. Independently from the physical mechanism involved, the detection of radiation introduces an intrinsic noise due to the fluctuations of the particles produced in the interaction with the material. In addition, there will be additional noises: Johnson noise, dark current noise, and readout noise. All noises are considered uncorrelated and are added in quadrature.

Radiation detectors are characterized by the *quantum efficiency* η, the ratio of the number of detected photons to the number of incident photons:

$$\eta = \frac{N_{detected}}{N_{incident}} \qquad (5.1)$$

The *detective quantum efficiency* (*DQE*) takes into account the signal-to-noise ratio of the measurement:

$$DQE = \frac{(SNR)_{out}^2}{(SNR)_{in}^2} = \frac{N_{out}}{N_{in}}, \qquad (5.2)$$

The detective quantum efficiency is always smaller than the quantum efficiency.

A detector at a telescope performs a sampling of the telescope image, after the distortion introduced by the instrument itself and, for ground-based facilities, also by the atmosphere. The detector contributes with its modulation transfer function that is related to its spatial resolution capability, i.e., to the size δ of its elements, the pixels. The corresponding Nyquist frequency is $(2\delta)^{-1}$. The pixel size of the detector array should provide an optimal sampling of astronomical image, ensuring that the FWHM of the Point Spread Function is larger than a few pixels.

The family of photon detectors is very extended; the references [2–5] provide detailed accounts of different types of detectors. The *photoemission devices* are based on the photoelectric effect, where a photon extracts an electron from a material. Photomultiplier tubes and microchannel plates are examples of photoemission detectors. The *photoabsorption devices* rely on the production of an electron by the absorption of a photon in the material, usually a semiconductor. The physical process involved in the detection is the *photovoltaic effect* and the *photoconductive effect*. The photovoltaic detectors use junctions of different materials with a depletion region; an example is given by photodiodes. The photoconductive detectors use single materials whose conductance is changed by the absorption of photons.

5.2 Photographic Plates

Photographic plates have been the first media to record astronomical processes independently from the ability and the experience of the human observer and have dominated astronomy for decades. Differently from the human eye, plates allow long integration times and have given to astronomy the capability to measure faint objects. The plates are made of AgBr crystals inside a gelatin matrix. AgBr is sensitive to light. The absorption of a photon promotes an electron in the conduction band; recombination is avoided using impurities that neutralize the silver ions to silver atoms to form a latent image. During the development, silver nuclei act as catalyzers to convert AgBr to silver, but the process is stopped and the fixing eliminates the residual AgBr, producing a negative image. The quantum efficiency of photographic plates is smaller than a few percent. The spectral response is limited to the blue side of the spectrum. The response of photographic plates is not linear.

5.3 Photomultiplier Tubes

The *Photomultiplier Tubes* (PMT) are photoemissive devices based on the photo-electric effect in vacuum [2–5]. A photocathode is kept at an high negative voltage; an incident photon extracts an electron that is accelerated toward a first electrode, a dynode, coated with a material that produces the extraction of secondary electrons. The electrons are accelerated toward a second dynode and so on. The system of n dynodes is biased with a ladder of high voltages and produces an amplification of the signal proportional to δ^n, where δ is the number of secondary electrons. The photomultipliers have a linear response.

The signal is measured by counting the electron pulses. The final pulse is very fast, with a duration of the order of a few nanoseconds, making the photomultipliers suitable detectors for the timing of fast events, such as the occultations. The quantum efficiency of photomultiplier tubes is of the order of 20 to 30 % (Fig. 5.1), much larger than the efficiency of the photographic plates, but with a maximum in the blue region as the plates. The main source of noise in photomultiplier tubes is due to the dark current that can be reduced by cooling the instrument.

The *Micro channel plate* (MCP) is a variation of the principle of the photomultiplier tube. A microchannel is a capillary tube with a diameter of a few microns internally coated with a photocathode material for electron multiplication; several microchannel capillaries are fused together in arrays of up to 10^6 channels and coupled to an anode. The configuration named *Multianode Microchannel Array* (MAMA) replaces the anode with two planes of crossed wires to be used for imaging.

Fig. 5.1 Quantum efficiency of a photomultiplier tube used at the photometer of the ESO 50-cm telescope, http://www.ls.eso.org/sci/facilities/lasilla/

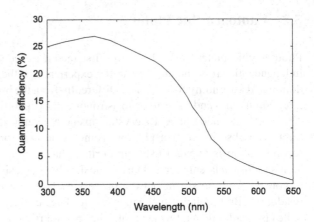

5.4 Photoconductors

The layout of a *photoconductor* [2–5] is shown in Fig. 5.2. The absorption of photons in semiconductors produces electron–hole pairs. The produced charges are collected by an applied electric field. The detection of visible photons is performed with intrinsic photoconductors made of pure materials. Their performances are defined by the energy E_g of the band gap between the valence band and the conduction band, that is of the order of a few eV for most materials. The cut-off wavelength, the largest wavelength that can be detected, is:

$$\lambda_c = \frac{hc}{E_g} \tag{5.3}$$

The band gaps for silicon and germanium are 1.1 and 1.8 eV, respectively. A flux of incident photons with frequency ν and power P produces an average photocurrent:

$$I_{ph} = \frac{e\eta P}{h\nu} \frac{\tau V}{d} \tag{5.4}$$

where η is the quantum efficiency, d the detector size, and V and τ the velocity and the recombination lifetime of electrons. The photoconductive gain is the quantity $G = \frac{\tau V}{d}$, the ratio of the electron lifetime to the transit time in the detector. The photocurrent shows a rms noise $\sqrt{4eI_{ph}GB_W}$, where B_W is the bandwidth of the measuring system. The response of a photoconductor is the ratio of the photocurrent I_{ph} to the power P, i.e., $S = \frac{e\eta G}{h\nu}$. Usually, photoconductors operate at low temperatures to reduce the contribution of charges generated by thermal processes.

Fig. 5.2 Layout of a photoconductor

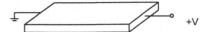

+V

5.5 Photodiodes

A *pn* junction contains a *depletion region* [2–5]. The absorption of radiation in the depletion region produces electron–hole pairs that are separated by the application of a bias field. The current produced by an incident flux of photons with frequency v and power P is:

$$I_{pd} = \frac{e\eta P}{hv} \tag{5.5}$$

The current has a rms noise $\sqrt{2eI_{pd}B_W}$. Compared to a photoconductor, a photodiode shows an unitary gain.

5.6 Superconducting Tunnel Junctions

The *Superconducting Tunnel Junctions* (STJs) are cryogenic detectors made of two superconducting films separated by a thin insulator [2–5]. The absorption of photons produces free charges by breaking the Cooper pairs in the superconductor. The number of produced charges is proportional to the energy of the absorbed photon. The band gap of the process is of the order of a few meV, three orders of magnitude smaller than that of semiconductor detectors. Thus, the STJ can be used for radiation ranging from the submillimeter to the X-rays wavelengths. The detectors provide intrinsic energy resolution, without the need of a dispersing element. The STJs used in the optical domain are based on Ta and have a resolving power of the order of a few tens at an energy of some eV. The *S-Cam* instrument[1] is a 10 × 12 array of Ta/Al STJs with a resolving power of 10.

5.7 Charge-Coupled Devices (CCDs)

The CCDs allow to perform *multiobject photometry*, securing the image of several objects at the same time, in particular allowing to measure the target object, the reference object, and the background at the same time. Charge-coupled devices (CCDs) are arrays of metal-oxide semiconductor (MOS) capacitors that have the capability to store electrons in a potential well [1]. The electronic circuitry needed to read the charges is integrated with the device. Details of the CCD and their applications in astronomy can be found in [1–5]. Photons are absorbed in the material through the photoelectric effect, governed by the silicon band gap, 1.1 eV, producing an electron-hole pair. The *photoelectrons* are stored in the depletion region. A suitable

---

[1] http://sci.esa.int/future-missions-office/41840-s-cam-3-optical-astronomy-with-a-stj-based-imaging-spectro-photometer/.

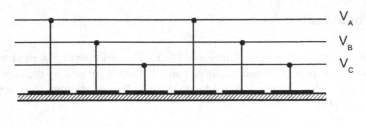

Fig. 5.3 Working principle of the three-phase CCD

configuration of applied voltages allows to move the charges from one pixel to the other. Each pixel is equipped with different gates, usually three, with voltages that can be separately tuned. One gate out of three is connected to a clock. The working principle of a three-phase CCD is shown in Fig. 5.3. The stored electrons are moved to close pixel by deepening the potential wells. The columns of the CCD are connected in parallel; thus, the charge transfer occurs at the same time along the rows.

The charges are transferred to an output *shift register*, a row of pixels not exposed to radiation. The whole process of the CCD readout requires from several seconds to minutes: CCDs are intrinsically slow detectors. The collected charge is transformed into a voltage. It is firstly amplified, with a typical *gain* of a few μV/electron [1]. Then, the amplified charge is converted into integer *analog-to-digital Units* (ADUs) by an analog-to-digital converter (ADC), with gains of the order of tens to hundreds of electrons per ADU [1]. The whole process is performed by the on-chip electronics with a *readout noise* of a few electrons per pixel.

Another source of noise, the *thermal noise*, produces a *dark current* according to [1]:

$$\frac{dN_d}{dt} \propto T^{\frac{3}{2}} e^{-\frac{E_g}{kT}} \tag{5.6}$$

The CCDs at professional telescopes are cooled at a temperature of $-100\,^\circ$C that decreases the dark counts by some orders of magnitude compared to the value at room temperature.

CCDs can operate as *front-illuminated* or *back-illuminated* devices [1]. The front-illuminated CCDs must have electrodes with open architecture to avoid the absorption of radiation on the green and blue side of the spectrum. The back-illuminated CCDs are sensitive to this region of the spectrum, but they must be thinned to avoid the recombination of the charges before the arrival to the depletion region. The sensitivity to photons on the red side of the spectrum is reduced.

Fig. 5.4 Quantum efficiency of the CCD (back-illuminated) in the BFOSC instrument at the Loiano telescope, Italy

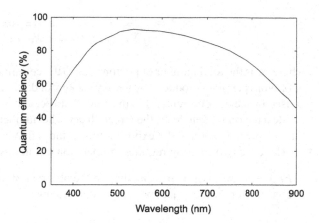

The quantum efficiency of CCDs is very high, up to 80 to 90 % over the most part of the visible spectrum. An example is shown in Fig. 5.4 for a back-illuminated CCD.[2]

The response of CCD is linear up to the saturation value of the ADC. Each pixel has a limited ability to store charges, up to the full well capacity, about 10^5 electrons [1].

The CCD shows an intrinsic noise due to the *shot noise* fluctuations of the number of photoelectrons n_{pe} that is related to the number of counts n in ADU through the gain G:

$$n = \frac{n_{pe}}{G} \tag{5.7}$$

In addition, there is the electronic *readout noise* σ_R^2. The two noises are added in quadrature:

$$\sigma^2 = \sigma_{pe}^2 + \sigma_R^2 = Gn + \sigma_R^2, \tag{5.8}$$

The performances of the CCD are described by the *CCD equation* [1]. The estimated flux of sources is related to the number of photoelectrons produced by the incident photons by the target, the sky background, the dark current, and the readout noise. Only the first three contributions follow the Poisson statistics that holds for the produced photoelectrons, not to the digital counts: the signal-to-noise ratio is $\sqrt{n_{pe}} = \sqrt{Gn}$. All noise sources are added in quadrature, with Poissonian terms contributing as $(\sqrt{n})^2$. The CCD equation [1] describes the signal-to-noise ratio (SNR) of the observations:

[2]http://www.bo.astro.it/loiano/TechPage/pagine/BfoscManualTechPage/BfoscManual.htm.

$$SNR = \frac{N_o}{\sqrt{N_o + n_{pixel}(N_{sky} + N_{dark} + N_R^2)}}, \qquad (5.9)$$

where N_o is the total number of photons due to the celestial object, N_{sky} is the number of photons per pixel produced by the sky background, N_{dark} is the number of electrons caused by the dark current, N_R^2 is the readout noise, and n_{pixel} is the number of pixels inside a region that encloses the target object. The number of photons N is related to the rate of arrival $\dot{N}$ and the exposure time t, through $N = \dot{N}t$.

There are two limiting regimes of operation of interest:

- *Photon noise limit*: Corresponding to the observation of bright objects, when the contribution of the object photons is dominating:

$$SNR = \sqrt{\dot{N}_o t} \qquad (5.10)$$

The error on the magnitude of bright objects is easily derived as $\Delta m = 1.087 \frac{1}{SNR}$.

- *Background limit*: Corresponding to the dominance of the sky contribution, the regime of large telescopes investigating faint sources:

$$SNR = \frac{\dot{N}_o \sqrt{t}}{\sqrt{n_{pixel} \dot{N}_{sky}}} \qquad (5.11)$$

The equation is inverted to estimate the exposure time. The CCD equation will be used in Chap. 13 to estimate the exposure time for the observation of targets with a given magnitude and a desired signal-to-noise ratio.

Problems

5.1 Discuss the difference between quantum efficiency and detective quantum efficiency.

5.2 Discuss the band gap in semiconductor and its impact.

5.3 Consider an 8-stage photomultiplier tube operating with a voltage of 1200 V with a secondary emission factor $\delta = 3.5$. Estimate the number of electrons produced for each photoelectron extracted at the photocathode.

5.4 The CCD is the most used detector for optical astronomy. Discuss the CCDs in comparison with the other families of detectors (photomultipliers, photoconductors, plates).

5.5 Discuss the working principle of a three-phase CCD.

References

1. Howell, S. B.: Handbook of CCD Astronomy, Cambridge University Press (2006)
2. Lèna, P. et al.: Observational Astrophysics. Springer-Verlag Berlin Heidelberg (2012)
3. McLean. I. S.: Electronic Imaging in Astronomy: Detectors and Instrumentation, Springer (2008)
4. Oswalt, T. D., McLean, I. S.: Planets, Stars and Stellar Systems. Volume I: Telescopes and Instrumentation. Springer (2013)
5. Oswalt, T. D., Bond, H. E.: Planets, Stars and Stellar Systems. Volume II: Astronomical Techniques, Software, and Data. Springer (2013)

Chapter 6
Optical Photometry

Optical photometry measures the apparent brightness of celestial objects in the optical region of the spectrum. The magnitudes and the colors of stars and galaxies allow to derive their physical properties and composition. Photometry is a form of very low-resolution spectroscopy, since it measures the spectra of sources in a few points, determined by the passbands of the filters. After presenting the order of magnitude of typical astronomical objects, the photometric systems used in optical astronomy are discussed. The choice of their band pass is dictated the astrophysical quantity or feature of interest, leading to hundreds of different systems. The techniques for securing photometric observations with CCDs are discussed. The presence of the atmosphere produces an extinction of the signal. The instrumental magnitudes extracted in the reduction of the photometric data must be transformed to a standard system.

6.1 Astrophysical Optical Sources: Fluxes

In this chapter, we will present the procedures to extract the magnitude of an object, according to:

$$m = m_0 - 2.5 \log \frac{F}{F_0} \qquad (6.1)$$

where m_0 and F_0 are the zero point magnitude and the reference flux. The magnitude values estimated with the techniques described below are *instrumental magnitudes*, related to the specific characteristics of the instrumentation used during the observation. The instrumental magnitudes must be transformed to a standard system and corrected for the atmospheric effects. The apparent magnitude (in the photometric V band described below) of some objects of interest is reported in Table 6.1 to provide an estimate of the orders of magnitude.

The power of photometric observations is demonstrated by the *color–magnitude diagrams*, the graph of the absolute magnitude in one band against the color, the

© Springer International Publishing Switzerland 2017
R. Poggiani, *Optical, Infrared and Radio Astronomy*,
UNITEXT for Physics, DOI 10.1007/978-3-319-44732-2_6

Table 6.1 Apparent
magnitudes of some
astronomical sources

Object	m_V
Sun	−26.8
Full moon	−12.6
Sirius	−1.5
M31	3.4
3C 273	12.8
Cepheid of HST key project	27–28
HST deep field exposure	29.5

Fig. 6.1 Color–magnitude
diagram of M67, after [6]

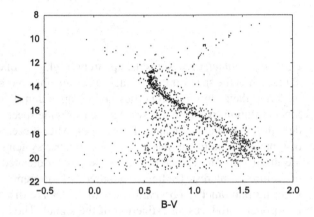

difference of the magnitudes in two bands. Generally, the diagram reports the V
magnitude and the $B - V$ color, described in the next section. The position on the
diagram provides information on the star type. An example of diagram is reported
in Fig. 6.1 for the open cluster M67.

6.2   Photometric Systems

The definition of a *photometric system* requires a detector, a set of *standard photometric filters* and a group of *standard stars* with known magnitude and colors [1].
Photometric filters are defined by a *transmission curve* $T(\lambda)$. Each filter is characterized by an *effective wavelength*:

$$\lambda_{eff} = \frac{\int \lambda F_\lambda T(\lambda)d\lambda}{\int F_\lambda T(\lambda)d\lambda}, \tag{6.2}$$

where F_λ is the monochromatic flux. The difference in the magnitudes measured
between two filters defined by the wavelengths λ_{lower} and λ_{upper} is named *color*

index or color:

$$color\ index = m(\lambda_{lower}) - m(\lambda_{upper}), \qquad (6.3)$$

To date, more than two hundreds photometric systems have been defined, only the most used ones will be presented. A complete listing is presented by [5] and is available online.[1] A critical review of the astronomical photometric systems has been presented by [1]. The photometric systems are classified into three families: *broadband* (filter widths of the order of some hundreds Å), *intermediate band* (widths in the range 70 to 400 Å), and *narrow band* (width smaller than 70 Å) systems. Broadband systems are widely used and will be discussed first. Since the overlap of the response functions is not negligible, they are replaced with intermediate band filters for some applications. Narrowband photometric systems are built around specific spectral lines, such as Hα and Hβ.

The first photometric system, the *visual band* m_{vis}, is defined by the response of the human eye. The peak frequency is in the region 515 to 550 nm, with a width of 82 to 106 nm. Other photometric bands have been defined using the photographic plates as detectors. The *photographic band* m_{pg} has a high transmission in the blue region, at about 400 nm, and a width of about 170 nm. The *photovisual band* is realized with the addition of a yellow filter to reproduce the visual band. The peak wavelength is about 550 nm, while the width is about 100 nm.

The broadband systems described in the following have been defined to match the previous observations and have selected materials that could allow an easy reproducibility at any observatory. The *Johnson UBV system* was defined using the blue-sensitive photomultiplier RCA 1P21 and a set of colored glass filters. The *V band* was defined to approximate the photovisual band, $V = m_{pv}$, the *B band* to approximate the photographic band and the *U band* to fill the region at the lower wavelengths. The UBV system has been extensively used for stellar astronomy. The color index $B - V$ measures the star temperature, while $U - B$ is sensitive to the Balmer discontinuity at 364 nm. The passbands of the UBV filters are compared with the normalized spectrum of Vega in Fig. 6.2.

The development of new photomultipliers, sensitive to the red part of the spectrum, has allowed the definition of the *Johnson RI system*, with the R_J band and I_J band, in the red and close infrared regions. Another system in the red region, based on the GaAs tube, is the *Cousins RI system*, with the R_C and I_C bands. The actual system is called *Johnson–Cousins UBVRI system*, combining the UBV section by Johnson and the RI section by Cousins. The complete Johnson system includes also the infrared photometric bands that will be discussed in Chap. 8. The transmission curves of the UBVRI system is reported in Fig. 6.3, left.

The Sloan Digital Sky Survey (SDSS) is performing photometric observations of stars and galaxies, with the potential to become the future reference, with an archive with about 10^8 objects. The *SDSS system* has defined the bands named u', g', r', i', z', loosely centered on the U, B, V, R, I filters, but with a smaller overlap. The transmission curves are reported in Fig. 6.3, right.

---

[1] http://ulisse.pd.astro.it/Astro/ADPS/.

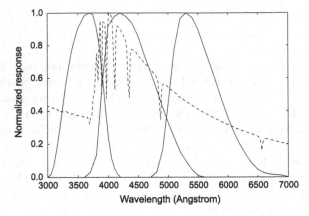

Fig. 6.2 Passbands of the U, B, and V filters compared with the normalized spectrum of Vega; data from [5]

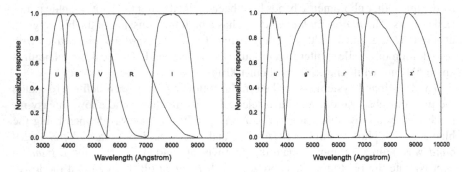

Fig. 6.3 Transmission curves of the UBVRI and SDSS systems; data from [5]

For completeness, we mention two other systems. The *HST WFPC2* has been defined for the Hubble Space Telescope. Among the several broadband filters, the bands F336, F439, F555, F675, and F814 are partially related to the UBVRI filters. The *Hypparcos–Tycho system* has been used for the photometry and astrometry of the hundreds of thousands stars investigated by the Hipparcos mission. The system defined the bands B_T, V_T, and H_P, with the first two bands very similar to the B and V filters of the UBVRI system.

The properties of the broadband systems described above are briefly summarized in Table 6.2 [1].

Due to the relevance of the UBVRI system, we report in Table 6.3 the flux corresponding to magnitude zero.

The intermediate band systems have been devised to overcome the overlap of the bands associated with the broadband systems. Among them, the most widely used is the *Stromgren system* with the *u, b, v, y* bands, well separated in response, useful for stellar temperature investigations [1] (Table 6.4). The transmission curves are reported in Fig. 6.4.

Table 6.2 Effective wavelength and band pass width of some widely used broad band systems: UBVRI, SDSS, Hipparcos, HST WFPC2; data from [1]

UBVRI			SDSS			Hipparcos			HST WFPC2	
	λ_{eff}(Å)	$\Delta\lambda$(Å)		λ_{eff}(Å)	$\Delta\lambda$(Å)		λ_{eff}(Å)	$\Delta\lambda$(Å)		λ_{eff}(Å)	$\Delta\lambda$(Å)
U	3663	650	u$'$	3596	570	H_P	5170	2300	F336	3448	340
B	4361	890	g$'$	4639	1280	B_T	4217	670	F439	4300	720
V	5448	840	r$'$	6122	1150	V_T	5272	1000	F555	5323	1550
R	6407	1580	i$'$	7439	1230				F675	6667	1230
I	7980	1540	z$'$	8896	1070				F814	7872	1460

Table 6.3 Flux at magnitude 0 for the UBVRI photometric system [1]

	Flux (10^{-30} W cm^2 Hz^{-1})
U	1790
B	4063
V	3636
R	3064
I	2416

Table 6.4 Effective wavelength and band pass width of the ubvy system [1]

	λ_{eff}(Å)	$\Delta\lambda$(Å)
u	3520	314
v	4100	170
b	4688	185
y	5480	226

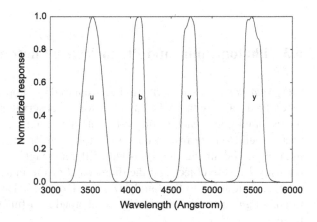

Fig. 6.4 Transmission curves of the Stromgren system; data from [5]

Fig. 6.5 Layout of a
photoelectric photometer

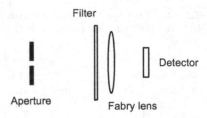

The bolometric magnitude, based on the emission of an object over the whole electromagnetic spectrum, is often determined by models that rely on the knowledge of the intensity in one or more bands. The approach is based on the *bolometric correction* (BC), the difference between the bolometric magnitude and the magnitude in one band, generally the V band:

$$BC = m_{bol} - V = M_{bol} - M_V \qquad (6.4)$$

The difference is zero for main sequence stars with a temperature of about 6500 K, i.e., F5 V stars.

The *AB magnitude system* [7] is a peculiar magnitude system, where the monochromatic flux F_ν is measured in erg s^{-1} cm^{-2} Hz^{-1}:

$$m(AB) = 2.5 \log F_\nu + 48.60 \qquad (6.5)$$

In this system, the spectrum of an object with constant flux per unit frequency has zero color.

6.3   Photographic and Photoelectric Photometry

The *photographic photometry* has been replaced by photometry with photomultiplier tubes firstly and with CCDs later. However, observatories have large archives of plates, often going back to the end of the nineteenth century. The archives are still used, to build long-term light curves and to search for possible transients that occurred in the past. The images of objects with different magnitudes have different sizes. The intensity I of the object and the diameter d of the image are related by empirical relations belonging to the family $D = a_1 + a_2 \log I$. The plates are scanned with a moving light spot whose intensity is recorded as a function of the position on the plate.

A *photoelectric photometer* (Fig. 6.5) measures the brightness of point-like sources using the photomultiplier tube, a detector with a single pixel. A physical pinhole, called *aperture*, is placed at the focus of the telescope to select the object. The concept of aperture is used also in the context of photometry with array detectors, where it becomes a software defined object. The size of the aperture must be larger than the seeing disk, but must not include an excessive sky background. The aperture is followed by a photometric filter. A Fabry lens is placed between the filter and the photomultiplier to produce a uniform image, to deal with the possible motion of the image or to tracking errors. The photometric observations with photoelectric photometers require the observation of the star, of a comparison star, and of the sky background that are measured sequentially by pointing the telescope in a cyclic procedure: sky-object-sky-reference.

The photoelectric photometers have dominated astronomical photometry until the advent of the CCDs. They are still used for high speed photometry of variable objects and occultations, due to their fast response.

6.4 Photometry with CCDs

Photometry requires *photometric nights*, with cloud-free skies. Even at very good sites, they are not the totality of the nights. Photometry with CCDs requires that data are calibrated for each observing night. In addition to science frames, it is necessary to secure different set of frames [3]:

- *Bias*: The bias frames (also called zero frames) are secured with closed shutter and a negligible integration time to measure the CCD noise. Bias frames must be secured for all night of observation. An alternative to securing bias frames is the use of *overscan*, a CCD region that is not illuminated; however, the bias frames estimate the noise over the whole frame.
- *Dark*: The dark frames are secured with closed shutter and with an integration time identical to the exposure time of the science object, to estimate the contribution of the dark current. The CCDs at professional observatories are cooled; thus, it is not necessary to acquire dark frames.
- *Flat*: The flat frames are secured using a source with uniform illumination to estimate the efficiency of the CCD pixel by pixel. It is necessary to secure flat frames for all filters that have been used during observations.

It is necessary to secure several bias and flat frames, ten at least of each kind, to improve the precision. The individual bias and flat frames are combined using the mean or the median pixel by pixel to build the *master bias* and the *master flat*. A bias frame is shown in Fig. 6.6, left; the image shows fluctuations about a mean value. A flat frame is shown in Fig. 6.6, right. Flat frames should be almost uniform as their name would suggest. The contrast of the image has been chosen to enhance the possible variation of the flat frame. The peculiar objects with a doughnut shape are the result of the diffraction on dust grains inside the instrumentation. Flat frames can be secured using different illumination sources. The *dome flats* are secured pointing the telescope out of focus toward a white screen inside the dome illuminated by a lamp with a continuous spectrum, with an exposure time set to fill about one half of the full well capacity. The *twilight sky flats* are secured at the twilight or at dawn pointing to

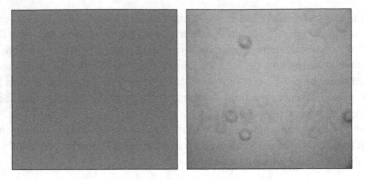

Fig. 6.6 Photometry with CCDs: bias frame (*left*) and flat frame (*right*)

some specific sky regions with a small luminosity gradient. The telescope is slightly shifted for each frame to remove the effects of the stars. The *sky flats* are built using the sky background in the science frames or performing targeted observations of empty sky regions. Long-time exposures are needed for twilight and sky flats. The choice of flat strategy is determined by the characteristics of the science target. Dome flats have a high signal-to-noise ratio, but the spectrum of the lamp is different from the spectrum of the sky, generally being redder. Sky flats have a color with a better match, but the signal-to-noise ratio is lower. In addition, the sky spectrum is affected by the variable contribution of the Moon that makes flat frames bluer. A practical rule of thumb is the choice of dome flats for bright objects, whose intrinsic color is dominant, and the choice of sky flats for fainter objects.

The reduction of raw imaging data requires several steps [3]:

1. subtraction of the bias frame from the object raw frame.
2. subtraction of the bias frame from the flat frame.
3. normalization of the bias subtracted flat frame to its mean or median value.
4. division of the bias subtracted science frame with the bias subtracted and normalized flat field

that are summarized by:

$$science = \frac{raw - bias}{(flat - bias)_{normalized}}, \tag{6.6}$$

The CCDs are sensitive to *cosmic rays* arriving during the exposure time. They appear as *hot pixels* with high or saturated counts; they can be removed by an average of the near pixel values or with σ-clipping techniques.

The statistical properties of the bias and flat frames are related to the CCD characteristics, the gain G and the readout noise RON. The histogram of the pixel counts of bias frames and of flat frames are both Gaussians, with standard deviations:

$$\sigma_{bias}(ADU) = \frac{RON}{G} \tag{6.7}$$

$$\sigma_{flat}(ADU) = \frac{\sqrt{G\bar{F}}}{G}, \tag{6.8}$$

where $\bar{F}$ is the mean value of the flat field. An estimation of the gain and of the readout noise of the CCD can be performed using a pair of bias frames, B_1 and B_2, and a pair of flat frames, F_1 and F_2:

$$G = \frac{(\bar{F}_1 + \bar{F}_2) - (\bar{B}_1 + \bar{B}_2)}{\sigma^2_{F_1 - F_2} - \sigma^2_{B_1 - B_2}} \tag{6.9}$$

$$RON = \frac{G\sigma_{B_1 - B_2}}{\sqrt{2}}, \tag{6.10}$$

The extraction of the magnitudes is performed with two main techniques [3, 8]: *aperture photometry* and *Point Spread Function fitting photometry*. In aperture photometry, the signal of the object is extracted within an aperture, usually circular, around the object, mimicking the approach of photoelectric photometry. In Point Spread Function fitting photometry the profile of each source is fitted with a suitable Point Spread Function, that is identical, apart the amplitude, for all stars in the same frame. The two techniques are interchangeable in uncrowded fields, with well separated sources. Point Spread Function fitting photometry is the only solution in crowded fields, such as globular clusters. The two techniques willbe discussed with reference to stellar images.

The first step in the photometric reduction is the determination of the centroid of star objects. Since a star profile is a function with a central peak, it is possible to provide an approximate estimation of the centroid by a search for local maxima. When the frame shows only isolated stars, as soon as approximate centroids are available, it is possible to use the *marginal sum method*. A region with a size of the order of a few times the FWHM of the frame Point Spread Function is chosen around each star and the marginal sums of the intensities along the rows and the columns are computed inside the box; the fit of each sum with a Gaussian provides an estimation of the centroid. Frames with a higher star density can be analyzed with the *image centroiding method*. The average of marginal sums in the two directions is computed, and the centroid is estimated by including only the points that are above the average and iterating the process if necessary. When the frame is crowded, it is necessary to directly move to the Point Spread Function fitting techniques.

In the method of aperture photometry, a region, called *aperture*, usually circular in shape, is drawn around the star to enclose the most part of the light (Fig. 6.7). The star flux I_{source} is estimated by summing the counts I_{ij} of the pixels (including the partial pixels) inside the aperture and subtracting the sky background contribution I_{sky} in the n_{pixel} of the aperture:

$$I_{source} = \sum I_{ij} - n_{pixel}I_{sky} \tag{6.11}$$

Fig. 6.7 Aperture
photometry: object aperture
and sky annulus

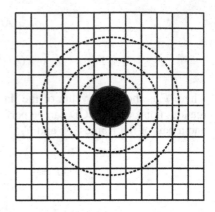

The radius of aperture should in principle contain the whole light from the star. Assuming a Gaussian Point Spread Function, a radius of $3 \times FWHM$ encloses almost 100 % of the star light. For bright source, large radii can be used without a degradation of the photometric error. For faint sources, small radii are a better choice, but they will not contain the whole light from the star. Starting from the star center, the increase in the aperture radius includes an increasing contribution of the star, but this contribution starts decreasing along the Point Spread Function wings and, for larger radii, the background will dominate. It has been demonstrated that the optimal aperture radius is $\sim 1 \cdot FWHM$ [3]. When dealing with bright stars, it is possible to use also larger aperture radii, of the order of a four or five times the FWHM value. Since the Point Spread Function of the stars in the frame is identical, independently from their brightness, the fraction of star light within a fixed radius is identical. Faint stars are measured with small apertures, while bright stars are measured with small and with large apertures. The difference between the values for small and large apertures for bright sources is named *aperture correction*.

The background is estimated locally around each star defining an annular region, the *sky annulus*, concentric to the star centroid and to the aperture defined above (Fig. 6.7). The inner radius of the annulus should be a few times the FWHM of the PSF to exclude the star radiation and thick enough to allow a good statistical estimation. Since the annulus could contain faint stars, cosmic ray hits etc., the mean of the pixels is not a robust estimator. The best estimator of the local background is the *mode*: $mode = 3 \times median - 2 \times mean$.

The method of Point Spread Function fitting is applied when stars overlap or several faint stars are present, relying on the uniformity of the Point Spread Function for all stars in the frame. It is possible to define an *empirical PSF* or an *analytical PSF*. The first approach has the advantage of making no assumption on the PSF profile. The Point Spread Function is built choosing some bright stars in the frame and modeling it with a combination of a Gaussians and some residuals. The second approach is easier from the computational point of view. Some examples of *analytical PSF* are:

- Gaussian function: $I = I_o \exp(-\frac{r^2}{2\sigma^2})$
- Moffat function, describing the seeing profile: $I = I_0 \frac{1}{(1+(\frac{r}{a})^2)^\beta}$

The fitting of the PSF is performed within a defined fitting radius. Generally, the data reduction softwares simultaneously fit several stars at a time.

The details of the reduction procedures for photometric observations, starting from raw data and the bias and flat frames, are presented in Chap. 14. The reduction of photometric data requires two important steps [3, 8]:

- correction of the effects of atmosphere, i.e. of the *extinction*;
- conversion to a standard photometric system.

A fundamental ingredient of the photometric systems is a set of *standard stars*, whose magnitudes and colors have been measured with high precision. Standard stars should be non-variable and should be available in any sky region targeted for observation. They should be distributed over large intervals of colors.

The standard stars for the UBVRI photometric system have been measured by [4]. The stars of the catalog have magnitudes in the range 13–15, with a typical error of the order of a few millimagnitudes, are uniformly distributed in right ascension and are accessible from both hemispheres. The Landolt stars can be observed with short exposure times (tens seconds) and good signal-to-noise ratios at telescopes with apertures of 1–3 m.

6.5 Atmospheric Extinction

Before discussing the atmospheric extinction, it is necessary to recall that the radiation is affected by the interstellar medium that attenuates and also scatters the radiation, producing the *interstellar extinction*. The *color excess* is defined as:

$$E(B - V) = (B - V) - (B - V)_0 \tag{6.12}$$

where the index 0 labels the intrinsic values. The interstellar reddening law relates the interstellar absorption A_V in the V band and the color excess $E(B - V)$ (Fig. 6.8).

The radiation collected by ground-based instrument is affected by the absorption and scattering of the incident radiation in the atmosphere. The path in the atmosphere is the product of the vertical depth of the atmosphere and of the *air mass X*, a function of the zenith angle z. For zenith angle smaller than 60°, the air mass is approximated by $X \simeq \sec z$. A more accurate estimation is given by [2]:

$$X = \sec z - 0.0018167(\sec z - 1) - 0.002875(\sec z - 1)^2 - 0.0008083(\sec z - 1)^3 \tag{6.13}$$

The air mass is 1 for objects at zenith and 2 at a zenith distance of 60 degrees. The combined effects of absorption and scattering of radiation in atmosphere produces the *extinction*. The extinction depends on the elevation of the object above the horizon.

Fig. 6.8 The interstellar reddening law [9]

Fig. 6.8 The interstellar reddening law [9]

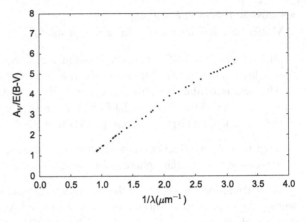

Fig. 6.9 Atmospheric extinction at the KPNO and CTIO observatories (data from http://iraf.noao.edu)

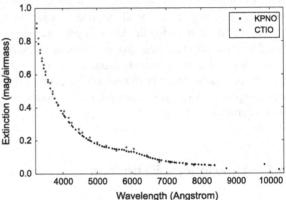

The object is brighter when it is higher in the sky, being brightest when at zenith. The extinction in the optical region measured at two astronomical sites is shown in Fig. 6.9. The Rayleigh scattering on air molecules and Mie scattering on aerosols are the dominant contribution to extinction in the optical region; the Rayleigh scattering dominates in the blue region, due to the steepest power dependence. The reported curves are only averages: the photometric observations require a careful estimation of the extinction for every night [2]. The extinction can be very different across a set of broadband filters. As an example, in the UBVRI system, the extinction is about 0.5 mag/air mass in the near UV, about 0.25 mag/air mass for the B band, slightly larger than 0.1 mag/air mass for the V, R, I bands.

Assuming to work with monochromatic magnitudes, the value m observed with ground-based instrumentation depends on the value m_0 measured outside the atmosphere and on the air mass X, according to the *Bouguer law* (Fig. 6.10):

$$m_0 = m - k(\lambda)X \tag{6.14}$$

Fig. 6.10 Bouguer law

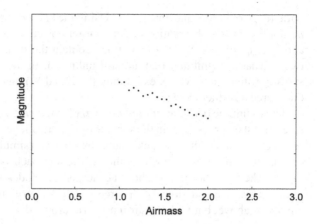

where k is the *extinction coefficient*. Assuming to monitor a star at different air masses, the extinction coefficient is estimated by the slope of the straight line of the graph of the observed magnitude m against the air mass.

6.6 Transformation to a Standard System

The observation of standards stars is of paramount importance: even if their observation could look as stealing the time of sources of interest, they are fundamental to extract values in physical units. Each observing session should include standard stars observed at different air masses and stars with different colors to account for color corrections. The transformation from the instrumental magnitudes to standard magnitudes is governed by a set of equations [8]. Focusing on the B and V bands, the link between the instrumental magnitudes, represented by lower case letters, and the magnitudes in a standard system, represented by uppercase letters, can be written as [3, 8]:

$$v = V + \alpha + \beta(B - V) + \gamma X \qquad (6.15)$$
$$b = B + \delta + \zeta(B - V) + \eta X \qquad (6.16)$$

The parameters that appear in the equation have different origins. The α and δ parameters are the zero points of the two bands; β and ζ are a measure of the difference between the observer band passes and the standard ones; γ and η are the first-order extinction coefficients; and X is the air mass. It is generally assumed that the air mass of an observation is precisely known.

When the transformation equations are applied to the observation of standard stars with known magnitudes and colors, it is possible to extract the value of the parameters using least square fitting. The sample of standard stars should cover a wide range of colors and air masses. Using the known parameters, the equations can be applied to the observations of the target objects, to extract the values of the B and V magnitudes. The two equations contain the quantity B-V, the difference of two standard magnitudes that are still unknown. Its value is searched by iteration, starting with a guess value, estimating the B and V magnitudes, and repeating until there are no further changes.

The technique of *differential photometry* is used when calibrated and non-variable reference stars are present in the same frame of the science target. Since the target, the reference stars, and the sky background are observed simultaneously, the atmospheric extinction is the same for all of them. The approach is particularly useful in the nights when the sky is not photometric and precludes the possibility of absolute photometry. The equations of differential photometry are the same transformation equations above, but with the air mass term dropped.

6.7 Astrometry with CCD

The CCDs are is currently used for relative astrometry, to determine the relative positions and the proper motion of stars with respect to reference object in the same frame [3]. The position of a star varies in time due to proper motion, the parallax effect, and so on. The astrometric observations involve securing photometric observations of the desired fields at different epochs, extracting each time the position of the stars and of the reference objects. The positions at all epochs are combined by the astrometric solution. Finally, the bias due to the average motion and parallax of the reference frame are subtracted from the observed positions. Astrometry is necessary any time a new transient is discovered. Photometric observations with high signal-to-noise ratio provide a precise estimation of the centroids of the target stars and of the reference star, using the techniques of the aperture photometry or Point Spread Function photometry. Back-illuminated CCDs are preferred for astrometric observations, since they have a better blue response and no subpixel variations caused by the penetration through the electrodes.

Problems

6.1 Estimate the flux per second per m^2 from stars with magnitude $V = 15$ and $V = 20$.

6.2 Discuss the techniques of aperture photometry and PSF fitting photometry.

6.3 Discuss the difference between absolute photometry and differential photometry.

6.4 Assume an average extinction of 0.14 mag/air mass in the V band. What is the fraction of radiation absorbed by the atmosphere for air masses of 1, 2, 2.5?

References

1. Bessell, M. S.: Standard Photometric Systems. ARAA **43**, 293 (2005)
2. Hiltner, W. A.: Astronomical Techniques, The University of Chicago Press (1962)
3. Howell, S. B.: Handbook of CCD Astronomy, Cambridge University Press (2006)
4. Landolt, A. U.: UBVRI photometric standard stars in the magnitude range 11.5-16.0 around the celestial equator. AJ **104**, 340 (1992)
5. Moro, D.; Munari, U.: The Asiago Database on Photometric Systems (ADPS). I. Census parameters for 167 photometric systems. A&AS **147**, 361 (2000)
6. Montgomery, K. A., Marschall, L. A., Janes, K. A.: AJ **106**, 181 (1993)
7. Oke, J. B.: Absolute Spectral Energy Distributions for White Dwarfs. ApJS **27** 21 (1974)
8. Oswalt, T. D., Bond, H. E.: Planets, Stars and Stellar Systems. Volume II: Astronomical Techniques, Software, and Data. Springer (2013)
9. Zombeck, M. V.: Handbook of Space Astronomy and Astrophysics. Cambridge, UK: Cambridge University Press (2007)

Chapter 7
Optical Spectroscopy

Astronomical spectroscopy is the main tool to investigate the physical properties of celestial sources and celestial environments, in particular the composition, temperature, and density. The most part of astronomical spectrographs in the optical domain is based on dispersing elements, such as the diffraction gratings. This chapter presents the discussion of the different dispersers and how they can be used in spectrographs. The spectral resolution is discussed for different situations. The spectroscopy with CCDs is presented.

7.1 Astrophysical Optical Sources: Spectra

This section presents some typical astronomical spectra to provide a reference to the reader, focusing on the spectra of stars, galaxies, and active galactic nuclei. The stellar spectra can be approximated by black bodies with some lines of atoms and molecules superimposed. The effective temperature of stars ranges from about 2500 to 50000 K; thus, their emission has a peak in the region around 1 μm. For example, the Sun has an effective temperature of about 5800 K, with an emission peak at about 500 nm. The stars are divided into different spectral classes (with decimal subclasses), labeled by the letters O, B, A, F, G, K, and M[1] and decimal subclasses. The main properties of the spectral classes are summarized in Table 7.1. The spectral type is an indicator of the star temperature that decreases when moving from the O to the M type.

---

[1] The sequence can be memorized using *Oh Be A Fine Girl/Guy Kiss Me*.

© Springer International Publishing Switzerland 2017
R. Poggiani, *Optical, Infrared and Radio Astronomy*,
UNITEXT for Physics, DOI 10.1007/978-3-319-44732-2_7

Table 7.1 Star spectral classes

Spectral class	Temperature (K)	Main lines
O	30,000–50,000	Ionized helium absorption
B	11,000–30,000	Neutral helium absorption
A	7,500–11,000	Strong neutral hydrogen
F	5,900–7,500	Strong ionized calcium, weak hydrogen and metals
G	5,200–5,900	Ionized calcium, neutral metals
K	3,900–5,200	Strong metal lines, weak molecular bands
M	2,500–3,900	Molecular bands (TiO)

Fig. 7.1 Spectra of stars of different spectral types and luminosity class V measured by [7]. Each spectrum is shifted vertically for clarity

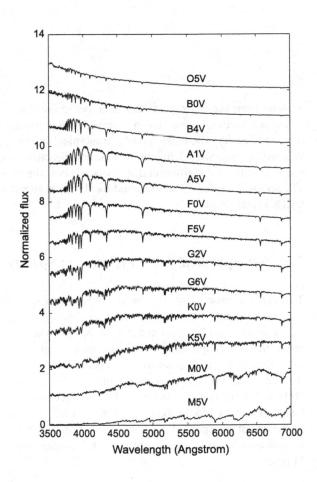

The spectra of several types with luminosity class V observed by [7] are reported in Fig. 7.1. The luminosity classes are: I (supergiants), II (bright giants), III (giants), IV (subgiants), and V (main sequence).

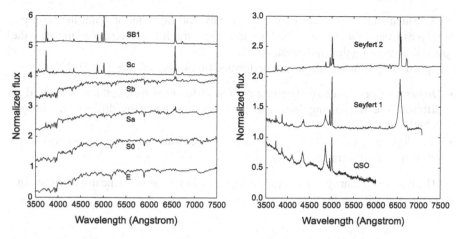

Fig. 7.2 Spectra of normal galaxies (*left*) and active galaxies (*right*). Each spectrum is shifted vertically for clarity

The template spectra of normal and active galaxies are reported in the left and right parts of Fig. 7.2. Normal galaxies include elliptical (E) galaxies, lenticular galaxies (S0), spiral galaxies (types Sa, Sb, and Sc), and starburst galaxy (type SB1) [2]. Galaxy spectra show a continuum produced by the component stars acting as black bodies that exhibit a large variety of temperatures. The elliptical galaxies show strong absorptions from old stellar populations. The spiral galaxies show emission lines, produced by young stellar populations, superimposed to the absorption features of the older stellar population. The evidence of the presence of a younger stellar population increases moving from spiral type Sa to Sc. The starburst galaxy, with bursts of star formation, shows the strongest emission lines. The templates of active galactic nuclei include Seyfert 1 and Seyfert 2 galaxies and a quasar (QSO).[2] The Seyfert 1 and 2 galaxies show a strong high-ionization emission lines. The type 1 Seyfert galaxies show two superposed families of emission lines, one with narrow widths of the order of several hundreds of km/s and one with large widths corresponding to the several thousands km/s; broad lines are observed only in permitted transitions. The type 2 Seyfert galaxies show only narrow lines. QSO or quasars spectra are similar to Seyfert spectra, but show weak narrow lines.

The observed spectrum S_o is the convolution of the intrinsic spectrum S_i of the source, the effect of the atmosphere S_a, and the combined instrumental response of the telescope optical elements and the spectrograph S_s:

$$S_o(\lambda) = \int S_i(\lambda) S_a(\lambda) S_s(\lambda) \tag{7.1}$$

---

[2]http://www.stsci.edu/hst/observatory/crds/cdbs_agn.html.

Astronomical spectroscopy extracts the physical information from the position, the intensity, and the shape of the lines in the spectra. It is customary to measure the spectral lines using the *equivalent width*.

Spectroscopic observations are performed by two techniques [1, 10–12]:

- *Dispersive spectroscopy*: The radiation at different wavelengths is dispersed in different directions using dispersing elements such as prisms, diffraction gratings, echelles, and grisms.
- *Non-dispersive spectroscopy*: The radiation spectrum is reconstructed using Fabry–Perot and Fourier transform spectrometers.

The most common spectroscopic instrumentation is based on the use of dispersing elements and will be discussed first.

7.2 Dispersive Optical Elements: Prisms, Gratings, Grisms, and Echelles

In the following, the spectroscopic instrumentation will be discussed as a function of the *spectral resolution*[1, 10–12]:

$$R = \frac{\lambda}{\delta\lambda} \tag{7.2}$$

where $\delta\lambda$ is the resolution element, i.e., the difference between two features that can be resolved. Resolution is considered low when R is smaller than a few hundreds, intermediate up to a few thousands and high above some thousands. Dispersing elements are evaluated in terms of the *angular dispersion* $\frac{d\theta}{d\lambda}$, the ability to separate different wavelengths.

Historically, the first elements used for dispersive spectroscopy have been the *prisms* (Fig. 7.3) that act as dispersers since their refraction index is a function of the wavelength, according to the Cauchy formula:

$$n(\lambda) = C_0 + \frac{C_2}{\lambda^2} + \frac{C_4}{\lambda^4} + \cdots . \tag{7.3}$$

Prisms are used in the condition of *minimum deviation*, achieved when the path inside the prism is parallel to its base. The minimum deviation wavelength corresponds to the central part of the spectrum. The angle of minimum deviation θ_m fulfills:

$$\sin\left(\frac{\theta_m + A}{2}\right) = n(\lambda)\sin\frac{A}{2} \tag{7.4}$$

where A is the aperture angle of the prism. The dispersion of the prism is not linear, since it behaves as λ^{-3}: blue light is diffracted more than red light. The resolution of

Fig. 7.3 Layout of a prism
(*left*)

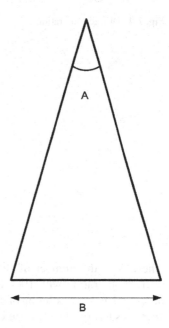

a prism is:

$$R = B|\frac{dn}{d\lambda}| \qquad (7.5)$$

where B is the prism base.

The standard dispersing element is a *diffraction grating*, a system with a periodic ruling defined by slits or mirrors whose axes are at a distance σ from each other. The grating can be used in *transmission* or in *reflection* [1, 10–12]

The incidence angle α and the diffraction angle θ (Fig. 7.4) fulfill the *grating equation*:

$$\sigma(\sin\theta \pm \sin\alpha) = m\lambda \qquad (7.6)$$

where the plus and minus signs correspond to reflection and transmission gratings, and m is an integer number, the *order* of diffraction. The angles α, θ share the same sign when they are on the same side of the grating normal. The angular dispersion of a grating for a fixed incidence angle is [12]:

$$\frac{d\theta}{d\lambda} = \frac{m}{\sigma\cos\theta} = \frac{\sin\alpha + \sin\theta}{\lambda\cos\theta} \qquad (7.7)$$

The dispersion depends only on the distance between slits and on the diffraction angle; it is approximately constant, since the cosine factor is slowly varying. High dispersion requires large orders and/or small ruling spacing. The intrinsic resolution of a diffraction grating is:

Fig. 7.4 Diffraction grating

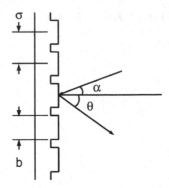

$$R = mN = \frac{mW}{\sigma} \tag{7.8}$$

where N is the number of slits and W the grating width. For example, a typical diffraction grating with 1000 lines/mm and a size of 100 mm operating at the first order will have a resolution of about 10^5. The evolution toward telescopes with larger apertures has been accompanied by the development of gratings with larger size.

The *free spectral range* defines the interval between the wavelengths λ_i and λ_j belonging to the adjacent orders and fulfilling $m\lambda_j = (m + 1)\lambda_i$:

$$\Delta\lambda_{FSR} = \lambda_j - \lambda_i = \frac{\lambda_i}{m} \tag{7.9}$$

Higher orders have a smaller free spectral range. The order superposition is particularly critical in spectrographs using CCD detectors that have a broadband response. The low orders are selected using blocking filters. At high orders, the free spectral range is small and orders are separated with cross-dispersers.

The basic diffraction grating directs the most part of light in the zero order that corresponds to pure reflection. In addition, there is the superposition between radiation diffracted into different orders [12]. The intensity of a wave diffracted by a grating with N facets of width b is proportional to the quantity:

$$\left(\frac{\sin Nh'}{N \sin h'}\right)^2 \left(\frac{\sin h}{h}\right)^2 \tag{7.10}$$

The first factor is the *interference function* that depends on the quantity $h' = \frac{\pi\sigma}{\lambda}(\sin\alpha + \sin\theta)$, one half of the phase difference between adjacent facets; the function has a maximum for $h' = m\pi$. The second factor is the *blaze function* that depends on the quantity $h = \frac{\pi b}{\lambda}(\sin\alpha + \sin\theta)$, the phase difference between the center and the side of the facet. The blaze function has a maximum when $h = 0$, i.e., when $\theta = -\alpha$, the pure reflection. The intensity pattern of a grating is given by the interference function modulated by the blaze function; thus, most of the light is

Fig. 7.5 Blazed grating (*left*) and echelle grating (*right*)

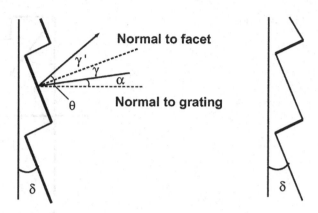

directed to the zero order. The efficiency in a specific order is increased by a suitable choice of the blaze function, using a *blazed grating*, a grating with a saw tooth profile defined by the blazing angle δ (Fig. 7.5, left).

The phase difference between the center and the side of a facet is $h = \frac{\pi\sigma\cos\delta}{\lambda}[\sin(\theta - \delta) + \sin(\alpha - \theta)]$. The new blaze function has a maximum for $h = 0$, i.e., for $\alpha + \theta = 2\delta$. The blaze condition corresponds to the specular reflection at the facet. The wavelength at the peak of the function is the *blaze wavelength*:

$$\lambda_b = \frac{2\sigma}{m} \sin \delta \cos \gamma \qquad (7.11)$$

where γ is the angle between the incident ray and the normal to the facet. The efficiency at different wavelengths is given by the blaze function. The efficiency of the grating decreases to 40 % of the peak value at the wavelengths:

$$\lambda_\pm = \frac{m\lambda_b}{m \mp \frac{1}{2}} \qquad (7.12)$$

For large orders, $\lambda_+ - \lambda_- \sim \frac{\lambda_b}{m}$.

The *echelle grating* is a blazed grating with a very high blazing angle (Fig. 7.5, right); thus, both incident and diffraction angles are large. Echelles have coarse rulings, of the order of tens to hundreds lines per mm and work at high orders, a few units at least. Echelle gratings have an high resolution, but at the cost of a smaller spectral range compared to traditional gratings. The echelle gratings are coupled to another dispersing element, the *cross disperser*, to distribute the different orders over the detector.

The *grism* (Fig. 7.6) is an hybrid element made of a diffraction grating operating in transmission coupled to a prism, to achieve a null total deflection of incident radiation. The diffraction grating is the real dispersing element; the prism has the role to realign the diffracted spectrum along the incident direction. The standard

Fig. 7.6 Grism

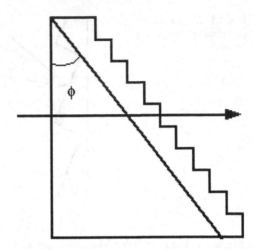

configuration consists of a prism with a right angle, with the grating mounted along the hypotenuse, with the facets parallel to the sides of the prism (Fig. 7.6).

Assuming an identical index of refraction n for the prism and the grating, the convergence along the optical axis is achieved for the wavelength:

$$\lambda_g = \frac{\sigma(n-1)}{m} \sin\phi \qquad (7.13)$$

where ϕ is the apex angle of the prism. The grism allows to perform spectroscopic observations without pointing again the telescope after the imaging of the field.

An alternative to standard diffraction gratings are the *Volume Phase Holographic* (VPH) gratings, periodic structures built by the modulation of the index of refraction inside the grating material. The incident radiation undergoes a Bragg diffraction in the grating layers.

7.3 Slitless Spectrographs

The simplest astronomical spectrograph can be realized by placing a prism (*objective prism*) in front of the objective [1, 12]. The configuration has the advantage of producing spectra of all sources in the field at the same time and is suitable for large-field Schmidt telescopes. Several historical photographic surveys have used the objective prism in the past, for example, the Henry Draper survey. The limit of slitless spectroscopy is the physical limit of the prism size that must be comparable with the telescope aperture. The difficulty in building large refracting optical elements with high optical quality limits the prism size of about 1 m. In addition, the spectra of different sources can overlap and the whole sky contributes to the background. The use of a dispersing element at the telescope entrance requires an offset in pointing that can be eliminated using a pair of prisms that provide the dispersion, but a zero net deviation of light.

7.4 Slit Spectrographs

The *slit spectroscopy* is the most used spectroscopic technique at professional telescopes [1, 12]. The most part of astronomical slit spectrographs consists of five basic elements (Fig. 7.7):

- *Slit*: placed at the telescope focus to select a point-like source or a part of an extended source.
- *Collimator*: to transform the radiation coming from the slit into a parallel beam.
- *Dispersing element*: in the following, we will assume that it is a diffraction grating with spacing σ operating at the order m.
- *Camera*: to focus the spectrum produced by the dispersing element onto the detector.
- *Detector*: usually, a CCD recording the intensity of light.

We will assume that the telescope has an aperture with diameter D_{tel} and focal length f_{tel}; the collimator has a diameter D_{col} and focal length f_{col}; and the camera has a diameter D_{cam} and a focal length f_{cam}. Losses are minimized when $\frac{f_{col}}{D_{col}} = \frac{f_{tel}}{D_{tel}}$. The slit, with physical width w_{slit}, selects a region of the sky with an angular size $\phi_{slit} = \frac{w_{slit}}{f_{tel}}$; spectrographs at telescopes offer a selection of slits with different widths to deal with different observing conditions. The presence of the dispersing element introduces the *anamorphic magnification* $r_{an} = \frac{d\theta}{d\alpha} = \frac{\cos\alpha}{\cos\theta}$. The dimension of the slit at the detector is $w_s = r_{an}\phi_{slit}\frac{D_{tel}}{D_{col}}f_{cam}$. The linear dispersion is $\frac{d\lambda}{dx} = \frac{1}{f_{cam}}\frac{d\lambda}{d\theta} = \frac{\sigma\cos\theta}{mf_{cam}}$. The product of the slit width at the detector and of the linear dispersion defines the resolution element:

Fig. 7.7 Layout of a slit spectrograph

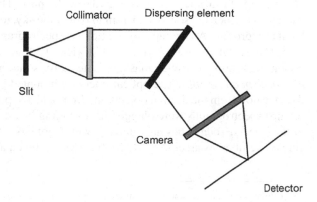

$$\delta\lambda = r_{an}\phi_{slit}\frac{D_{tel}}{D_{col}}\frac{\sigma\cos\theta}{m} \tag{7.14}$$

Finally, the resolution of a slit spectrograph in the *slit limited regime* is:

$$R = \frac{\lambda}{r_{an}\phi_{slit}}\frac{D_{col}}{D_{tel}}\frac{m}{\sigma\cos\theta} \tag{7.15}$$

The resolution is a function of the telescope aperture and of the collimator diameter: Collimator and cameras must be designed again when spectrographs are moved to a different telescope. The resolution increases by narrowing the slit width. However, for ground-based telescopes, the slit width must be matched to the dimension of the seeing disk. When spectrographs are installed on space-based telescopes, the slit width is matched to the size of the Airy disk, $\phi_{slit} = 1.22\frac{\lambda}{D_{tel}}$. The resolution of a spectrograph in the *diffraction limited regime* does not depend on the telescope diameter:

$$R = \frac{mD_{col}}{1.22\sigma\cos\alpha} \tag{7.16}$$

The echelle-based spectrographs [1, 12] work with multiple orders at the same time, oppositely to the operation of standard slit spectrographs that investigate a single order at a time. Each order spans a wavelength interval $\Delta\lambda = \frac{\lambda^2}{2\sigma\sin\theta}$ with an angular dispersion $\Delta\theta = \frac{\lambda}{\sigma\cos\theta}$. The lower orders have spectra that can extend beyond the size of the detector.

7.5 Advanced Spectrographs

The *multiobject spectrographs* are used in sky surveys to produce the spectra of all sources in the field of view at the same time [1, 10–12]. The *multislit spectrographs* use a slit mask with several slits matching the positions of the objects in the field. The observation of a different field requires a new slit mask. The single slits must be large enough to include the source and a small part of the sky, to allow an estimation of the local background. An example of multiobject spectrograph is the GMOS instrument at the Gemini Observatory,[3] with a field of view of 5.5 square arc min and a capability of some tens slits in a single mask. An alternative solution is the use of fibers in the *fiber-fed spectrographs*. The optical fibers are mounted in the focal plane to accept the radiation from the desired objects and from selected parts of the sky. The fibers at the first spectrographs were plugged in supporting boards with holes at the position of the objects. Today, they are managed with robotic positioning systems. The other side of the fibers is assembled far from the telescope on an optical bench.

---

[3]https://www.gemini.edu/sciops/instruments/gmos/multiobject-spectroscopy.

The *integral field spectrometers* provide the simultaneous acquisition of the spectra of different parts of an extended object. The data are stored as data cubes, whose coordinates are the spatial coordinates and the wavelength. The technique produces the same result of a progressive acquisition of spectra with a scanning slit, but is able to produce the data cube in a single step.

7.6   Spectroscopy with CCDs

The final detector of an optical spectrograph chain is a CCD [6, 10, 11]. A raw spectrum secured with a CCD is shown in Fig. 7.8. The directions along the array are the *dispersion direction*, defining the spectrum in wavelength, and the *spatial direction*. Both directions are measured in pixels in the raw spectrum frame. The spectrum of the target is contained inside the vertical strip close to the center of the frame, since the dispersion axis is vertical. Emission lines appear as slightly larger spots, while the opposite is true for absorption lines. The conversion of the pixel scale of the dispersion to a wavelength scale is performed by securing the spectrum of a spectral lamp with a large number of narrow lines, together with the science spectrum.

Spectroscopic observations require several calibration frames. The first set is the set of *bias* frames (Fig. 7.9, left), in analogy with the bias discussed in the context of photometry. The flat frames (Fig. 7.9, right) are obtained by using a featureless lamp whose spectrum must be removed. The details of the reduction procedure are discussed in Chap. 14.

Fig. 7.8 Spectroscopy with CCDs: raw spectral frame

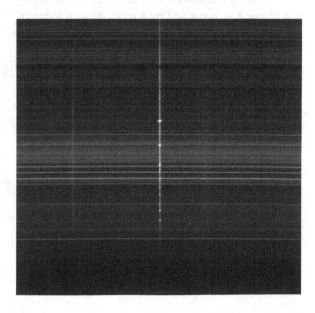

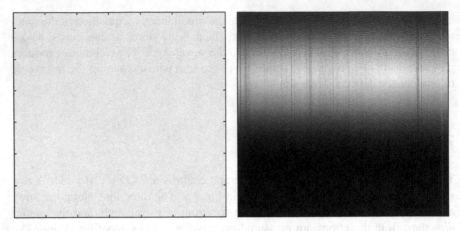

Fig. 7.9 Spectroscopy with CCDs: bias frame (left) and flat frame (right)

The data are calibrated in flux using spectrophotometric standard stars [3–5, 8, 9].

7.7   Non-dispersive Spectroscopy

The *Fabry–Perot etalon* (Fig. 7.10) achieves an higher resolution over a narrow band, compared to grating and to echelles, without using dispersing elements [1, 10–12]. The incoming radiation is split into beams using a pair of partially reflecting mirrors at a distance d, separated by a medium with index of refraction n, generally a gas. The instrument properties can be tuned by varying the distance of the plates or the refraction index of gas by tuning its pressure.

The beams then undergo interference and are partially transmitted with a normalized intensity:

Fig. 7.10 Layout of a Fabry–Perot etalon

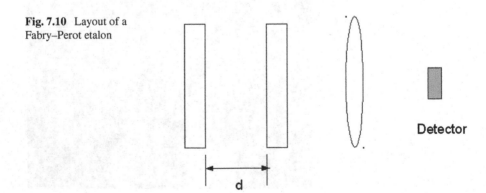

$$T_t = \frac{\left(\frac{T}{1-R}\right)^2}{1 + \frac{4R}{(1-R)^2} \sin^2 \frac{\delta}{2}} \qquad (7.17)$$

where T, R are the fractions of transmitted and reflected energy; $\delta = \frac{2\pi}{\lambda} 2nd \cos \alpha$ is the phase difference between consecutive beams; and α is the incident angle of the radiation. The transmission is maximum when $\delta = 2m\pi$, thus for $m\lambda = 2nd \cos \alpha$. The free spectral range of the Fabry–Perot interferometer is $\frac{\lambda^2}{2nd \cos \alpha}$. A predisperser is required to select the required order. The resolution of a Fabry–Perot etalon is:

$$R = \frac{2F_F nd}{\lambda} \qquad (7.18)$$

where $F_F = \frac{\pi \sqrt{R}}{1-R}$ is the *finesse*.

The *Fourier transform spectrometer* [1, 10–12] is a Michelson interferometer operating in scanning mode, whose output signal is processed to extract a spectrum (Fig. 7.11). The instruments divides the incoming beam into two halves using a beam splitter. The light reflected by the end mirrors is divided again at the beam splitter and recombined and directed toward the photon detector.

The fraction of the incident beam that appears in the output beam is related to the optical path difference Δ between the two arms of the interferometer:

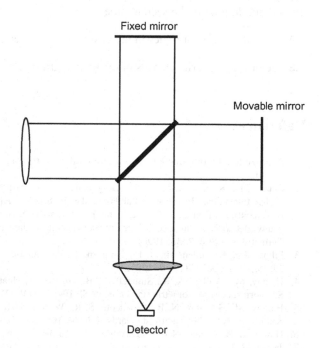

Fig. 7.11 Michelson interferometer

Fixed mirror

Movable mirror

Detector

$$\frac{I_{out}}{I_{in}} = \frac{1}{2}\left[1 + \cos(2k\Delta)\right] \tag{7.19}$$

The output flux measured for incoming radiation with a spectrum $S_{in}(k)$ is the Fourier cosine transform of the spectrum.

Problems

7.1 Discuss the main components of an astronomical spectrograph and their function.

7.2 Compare the performances of prisms and diffraction gratings as dispersing elements.

7.3 Compare the performances of unblazed and blazed diffraction gratings.

7.4 Discuss the role of the slit in astronomical spectrographs and its impact on the resolving power.

7.5 Discuss the technical solutions for high-resolution spectroscopy.

7.6 Estimate the number of rulings for a diffraction grating that is able to resolve the sodium doublet in the second order.

7.7 Describe the main issues in the design of a slit spectrograph.

7.8 Compare the performances of diffraction gratings and Fabry–Perot etalons.

References

1. Appenzeller, I.: Introduction to Astronomical Spectroscopy, Cambridge University Press (2013)
2. Calzetti, D., Kinney, A. L., Storchi-Bergmann, T.: Dust extinction of the stellar continua in starburst galaxies: The ultraviolet and optical extinction law. ApJ **429**, 582 (1994); Kinney, A. L., Calzetti, D., Bohlin, R. C., McQuade, K., Storchi-Bergmann, T., Schmitt, H. R.: Template Ultraviolet to Near-Infrared Spectra of Star-forming Galaxies and Their Application to K-Corrections. ApJ **467**, 38 (1996)
3. Filippenko, A., Greenstein, J. L.: Faint spectrophotometric standard stars for large optical telescopes. I. PASP **96**, 530 (1984)
4. Hamuy, M', Walker, A. R., Suntzeff, N. B., Gigoux, P., Heathcote, S. R., Phillips, M. M.: Southern spectrophotometric standards. PASP **104**, 533 (1992)
5. Hamuy, M., Suntzeff, N. B., Heathcote, S. R., Walker, A. R., Gigoux, P., Phillips, M. M.: Southern spectrophotometric standards, 2. PASP **106**, 566–589 (1994)
6. Howell, S. B.: Handbook of CCD Astronomy, Cambridge University Press (2006)
7. Jacoby G.H., Hunter D.A., Christian C.A.: A library of stellar spectra. ApJSS **56**, 257 (1984)

8. Massey, P., Strobel, K., Barnes, J. V., Anderson, E.: Spectrophotometric standards. ApJ **328**, 315 (1988)
9. Oke, J. B.: Faint spectrophotometric standard stars. AJ **99**, 1621 (1990)
10. Oswalt, T. D., McLean, I. S.: Planets, Stars and Stellar Systems. Volume I: Telescopes and Instrumentation. Springer (2013)
11. Oswalt, T. D., Bond, H. E.: Planets, Stars and Stellar Systems. Volume II: Astronomical Techniques, Software, and Data. Springer (2013)
12. Schroeder, D.: Astronomical Optics. Academic Press (1999)

Part III
The Low Energy Side of Classical Astronomy

Chapter 8
Infrared Astronomy

Infrared astronomy [1] investigates the region of the electromagnetic spectrum from 1 to about 300 μm and probes the emission of cold objects and of regions, like the center of our galaxy, that are not observable by optical instruments since they are inside dust. This chapter presents the telescopes and the detectors used in the different regions of the infrared, together with the observational techniques needed to deal with the large background.

8.1 Astrophysical Infrared Sources: Fluxes and Spectra

Infrared astronomy [1] is divided into different regions, according to the detection and observational techniques: the *near infrared*, between 1 and 5 μm; the *mid-infrared*, from 5 to 25 μm; and the *far infrared*, from 25 to 350 μm. The upper limit is defined by technological considerations, since it is the wavelength where the superheterodyne technique, typical of the radio astronomy, replaces the standard combination of telescope and detector. The black body radiation of the atmosphere and of the telescope itself become the dominant source of background between 2 and 3 μm, achieving the maximum contribution at about 10 μm. The signal of astronomical sources must be estimated by chopping, alternating between observing the region with the source and a region of blank sky. The high background forces the readout of the detectors at high rates, of the order of hundreds Hz, to avoid saturation. Differently from the optical observations, infrared observations are dedicated to reduce the contribution of the background against the faint signals of interest.

The infrared sky is very different from the optical sky [1]. According to the Wien law, the infrared spectral region investigates the sources with a black body temperature lower than in the optical region. Cool objects, with effective temperatures below 3000 K, can be observed in the near-infrared region (0.7 to 5 μm), where dust is almost transparent. The galactic center has been observed in the infrared: The extinction in its direction is about 2 magnitudes at 2.2 μm. Dust heated by the stellar radiation can be observed in the mid-infrared region (5 to 25 μm). The far-infrared region

© Springer International Publishing Switzerland 2017   109
R. Poggiani, *Optical, Infrared and Radio Astronomy*,
UNITEXT for Physics, DOI 10.1007/978-3-319-44732-2_8

Table 8.1 Flux of a sample of infrared sources at different wavelengths; data from [6]

Object	Wavelength (μm)	Flux (Jy)
Stars	20	$1–10^3$
H II regions	20	$1–10^3$
Molecular clouds	100	$10^4–10^5$
Sgr A	20	2600
Active Galactic Nuclei	20	6–30

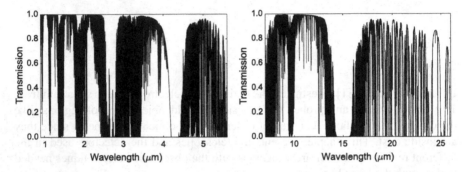

Fig. 8.1 Transparency of sky at the Gemini Observatory: near infrared (*left*) and mid-infrared (*right*), data available at http://www.gemini.edu/sciops/telescopes-and-sites/observing-condition-constraints/ir-transmission-spectra

(25 to 350 μm) allows to investigate the emission of cold dust and the cool molecular clouds. The molecular vibrational and rotational transitions of CO_2, CO, H_2O, H_2, silicates, and polycyclic aromatic hydrocarbon (PAH) are in the infrared region.

The flux of bright infrared sources of different classes is reported in Table 8.1 for reference.

The transparency of the atmosphere in the infrared spectral region is shown in Fig. 8.1. The limited regions where the transmission is higher have determined the definition of the infrared photometric filters. The atmosphere transparency in the infrared is affected by the strong absorption bands of water vapor and other components. The best sites for infrared observatories are at high altitudes, on dry mountains such as Mauna Kea in Hawaii or in Antarctica. Ground-based infrared astronomy explores the range between 1 and about 10 μm. Infrared observatories installed on board of aircrafts, such as SOFIA, operate at an altitude larger than 10 km and experience a smaller absorption by the atmosphere. The observations above some tens μm are performed with satellite-based instrumentation.

8.2   Infrared Telescopes

Optical telescopes can be used also in the very near-infrared region [1]. As discussed in Chap. 1, the thermal radiation of the sky and of the internal components of the telescopes produce a large background to infrared observations above about 2.3 μm. The telescopes operating in the region of thermal infrared are designed by minimizing the amount of radiating elements and surfaces. The secondary mirror is undersized and is supported by structures that are completely covered by it. The primary and secondary mirrors are coated with gold that shows a high reflectivity, but a small infrared emissivity (about 0.006 to 0.016 at 10 μm). The cooling of the internal components of the telescope is not viable for ground-based facilities, due to the unavoidable condensation of water vapor from the surrounding atmosphere. A field lens in the focal plane of the telescope produce an image on a *cold stop*, internal to the instrumentation. The seeing in the infrared region is better than in the visible one, since the Fried parameter is proportional to $\lambda^{\frac{6}{5}}$. The isoplanatic angle is proportional to the Fried parameter; thus, it is larger in the infrared. However, there is still the necessity to use laser guide stars. The telescope is moved between different locations, in the technique of *nodding*, to remove the contribution of possible inhomogeneities in the ambient temperature. The region of thermal infrared around 10 μm is dominated by the atmospheric background acting as a room temperature black body and showing a large variability. The subtraction of the background is achieved by modulating the sky emission with the *chopping* of the secondary mirror at a frequency of a few Hz. The two images in the focal plane are subtracted to remove the lower frequency noise.

Space-based infrared telescopes are not affected by atmospheric emission and are limited only by the astronomical infrared background. The operation in space allows the cooling of the complete telescope assembly. Cooling can be passive or active. Passive cooling uses the low temperature of space and a combination of Sun blocking and radiation shielding to achieve temperatures of a few tens Kelvin, providing an extended lifetime of the mission. Active cooling is achieved by placing the detectors and the optical assemblies inside a cryostat cooled by cryogenic liquids, nitrogen, or helium. The cryostats can be nested, enclosing an inner liquid helium cryostat inside a liquid nitrogen cryostat: The duration of the inner cryogen is prolonged. The duration of missions using active cooling is limited by the supply of cryogenic liquids.

8.3   Infrared Detectors

Infrared detectors show several sources of noise intrinsic to their operation [2–5]. Infrared astronomy is dominated by very large backgrounds, for example, the contribution of the atmosphere and the telescope itself for ground-based instruments. The *noise-equivalent power* (NEP) is the minimum detectable power, i.e., the power that

Material	λ_c (μm)
$Hg_xCd_{1-x}Te$	0.8–20
InSb	5.5
Si:Ga	17
Si:As	23
Ge:Ga	115

Table 8.2 Cut-off wavelength of materials for infrared detectors [1]

produces an unitary output signal-to-noise ratio. The detective quantum efficiency is given by:

$$DQE = \left(\frac{NEP_{BLIP}}{NEP}\right)^2 \qquad (8.1)$$

where NEP_{BLIP} is the NEP of the *Background Limited Infrared Performance* (BLIP):

$$NEP_{BLIP} = \left(\frac{4P_Bhc}{\lambda}\right)^{\frac{1}{2}} \qquad (8.2)$$

where P_B is the background power.

Infrared astronomy uses photon detectors belonging to the families of photoconductors and impurity band conduction detectors. The *photoconductors* for infrared astronomy have a band gap E_g smaller than the gap of the silicon used for CCDs, whose cut-off wavelength is about 1.1 μm. The detectors must be cooled to reduce the contribution of thermally excited electrons, with a lower operating temperature for the materials with the smaller band gaps. The properties of common materials for infrared detection are reported in Table 8.2.

Two intrinsic materials have a band gap suitable for infrared astronomy [1–5]. Indium antimonide (InSb) has a cutoff wavelength at about 5 μm. Mercury cadmium telluride (HgCdTe) has a band gap energy that can be tuned by varying the relative fraction of cadmium, from about 1 to some tens μm. Silicon and germanium with the addition of dopants show additional impurity levels between the valence and the conduction bands, producing *extrinsic photoconductivity* [1–5]. The absorption coefficient is proportional to the density of the impurities and to the photoionization cross section, that is, of the order of 10^{-15} cm^2. The concentration of the impurities cannot be increased arbitrarily, to avoid the formation of an impurity band and a large increase in the dark current. An exotic variation of the photoconductor concept is the *stressed detector*. The cutoff wavelength of Ge:Ga can extended to higher wavelengths (up to 200 μm) by applying an uniaxial stress. The solution has been adopted for the MIPS instrument of Spitzer.

The most used detector in the region of a few tens μm is the *impurity band conduction* (IBC) or *blocked impurity band* (BIB) detector (Fig. 8.2) [1–5]. The device has an infrared absorbing layer with a doping level high enough to produce

Fig. 8.2 Layout of an
impurity band conduction
detector

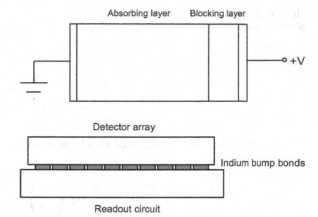

Fig. 8.3 Layout of an
infrared array

the impurity band. The absorption probability of photons is increased compared to a standard extrinsic photoconductor, with a reduction in the thickness of the detector. The large dark current is blocked by an undoped blocking layer.

Infrared arrays have been developed for the near- and mid-infrared regions of the spectrum [1–5]. The working principle is remarkably different from the operation of CCDs. Infrared arrays (Fig. 8.3) are assembled from a detector layer and a readout layer. The former is an array of infrared sensitive elements (HgCdTe, InSb, etc.). The latter is a pure electronic assembly, called multiplexer. The two layers are assembled together, pixel by pixel, using single indium bumps. The welding of a layer onto the other is achieved by mechanical pressure. Detected photons in each pixel are transformed into electrons that are discharged into the pixel capacitance. The well capacity of each pixel must be high enough to avoid saturation. The large background forces the readout and the resetting of the detector at rates of the order of tens Hz at least. There is no equivalent of the mechanical shutter used to control exposures with CCDs. Another difference is the readout of the signal that is performed on all detector pixels at the same time. Each pixel is matched by its own dedicated electronics for charge collection and resetting. An example of infrared array is the NICMOS instrument of Hubble Space Telescope, based on HgCdTe elements.[1]

The *bolometers* are thermal sensors (Fig. 8.4). The absorption of incident radiation with power P, regardless of its wavelength, produces a temperature increase ΔT that is related to the intensity of the radiation [1–5]. The absorber, with heat capacity C, is coupled to a thermal bath at temperature T_0 through a thermal conductance G. The bolometer can be modeled as an RC circuit.

The energy balance of the system is described by:

$$C\frac{\Delta T}{dt} + G\Delta T = P \tag{8.3}$$

[1]http://www.stsci.edu/hst/nicmos/.

Fig. 8.4 Working principle
of a bolometer

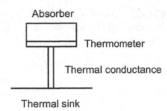

The temperature rise is measured by a thermometer coupled to the absorber. The temperature evolves as:

$$\Delta T = \frac{P}{G} \exp\left(-\frac{t}{\tau}\right) \tag{8.4}$$

where $\tau = \frac{C}{G}$ is the thermal time constant of the system. The thermal capacity of the absorber should be as small as possible. The limit of the bolometers is given by the thermal fluctuations occurring in the link to the thermal bath. The noise-equivalent power is:

$$NEP = \frac{\sqrt{4k_B T^2 G}}{\eta} \tag{8.5}$$

where η is the quantum efficiency. The thermal conductivity of the link should be minimized. The noise is reduced by operating the bolometers at cryogenic temperatures, generally below 1 K. The signal of bolometers is processed with low-noise JFET amplifiers. Since their operation temperature is several tens K, the amplifiers must be located in separate sections of the cryostat. A strong effort has been devoted to the development of the thermometers. Often they are realized by doping part of the semiconductor material or using neutron transmutation doping. Another solution is the use of superconducting films in the *transition edge sensors* (TES), whose resistance is strongly dependent on the temperature close to the superconducting transition. An example of application of bolometers for infrared astronomy is the PACS instrument of the Herschel observatory,[2] with two arrays of silicon bolometers.

8.4 Infrared Imaging

Infrared photometry requires that the detector observes only a small region of the sky around the science targets to reduce the contribution of the background. The detectors are cooled to low temperatures to reduce the thermal current and mounted inside cooled enclosures. The photometric filters must be cooled to reduce their emission.

---

[2]https://herschel.jpl.nasa.gov/pacsInstrument.shtml.

Table 8.3 Effective wavelength, filter width, flux at magnitude 0, of the JHKLM filters

Band	λ_{eff} (nm)	$\Delta\lambda$ (nm)	Flux $(10^{-30}\ \mathrm{W\ cm^2\ Hz^{-1}})$
J	1220	213	1589
H	1630	307	1020
K	2190	39	640
L	3450	472	285
M	4750	460	154

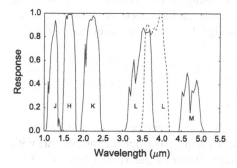 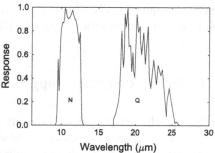

Fig. 8.5 Passbands of infrared photometric filters

The standard photometric system used in the infrared is the *Johnson-Glass JHKLMNQ* system [1]. The *JHKLM* filters are matched to the response of the InSb detectors and are centered at about 1.20, 1.65, 2.20, 3.5, and 5.0 µm, in the near infrared. The *NQ* bands are defined over the response of photoconductive or bolometric detectors and are centered at 10 and 20 µm, the region of mid-infrared. The response curves of *JHKLMNQ* filters are literally carved inside the regions of high atmospheric transmission. Some alternative filters have been defined to deal with the atmospheric absorption bands, by narrowing the bandwidth of the standard filters and making the sides of the response curve steeper. The *L* filter is sometimes replaced by the *L'* filter. The main characteristics of the filters are reported in Table 8.3. The passband of the JHKLM filters is shown in Fig. 8.5.

Absolute calibrations of photometric observations in the near infrared can be performed using laboratory black bodies. However, the calibrator has an intensity differing by several orders of magnitudes from the observed flux of the target sources. The use of the Sun as a calibrator requires the knowledge of its spectral distribution and has been used for the calibration of some infrared filters. The colors $V-J$, $V-K$, and $V-L$ of the Sun have been estimated using a set of stars sharing its spectral class. Starting from the apparent V magnitude of the Sun, extracted as a weighted mean of different observations, the J, K, L apparent magnitude were estimated. The spectral distribution of the Sun was used to compute the flux of a star with zero magnitude in the band of the desired filters. Finally, the theoretical spectra of stars with precise measurements in the optical region can be extrapolated in the infrared region. Theoretical models of stellar spectra are used also in the mid-infrared. Direct methods use a bright calibrator source, the Sun or Mars, following the same method described above for the calibration in the near-infrared region. The magnitude of Mars has been measured by comparison with a laboratory black body on Earth and with a calibrated radiometer in space. The flux of Mars was compared to the flux of a set of comparison stars.

Photometric observations can be secured using photometers or array detectors. Photometers use the chopping technique described above, where the oscillating secondary defines two regions in the sky, one enclosing the target and another containing only the empty sky. The signal is fed to a *lock-in*, an instrument synchronized to the chopper that performs the difference of the signals with and without the source.

8.5   Infrared Spectroscopy

Spectrographs operating in the near infrared [1–5] are similar to the optical spectrographs with dispersing elements discussed in Chap. 7. The optical components with good transmission efficiency in the infrared are made of quartz, sapphire, and ZnS. The cryogenic operation required for the reduction of thermal noise forces the use of cold slits. Some examples on infrared astronomical spectrographs are: the NIRSPEC instrument at Keck II,[3] a cross-dispersed echelle spectrograph with an InSb array for the region from 0.95 to 5.5 μ, equipped also with a slit viewing HgCdTe array for imaging; the Near-Infrared Camera and Multi-Object Spectrometer (NICMOS) of HST, in addition to provide imaging, is used for slitless grism spectroscopy in the infrared region from 0.8 to 2.5 μm.

8.6   Ground-Based and Space-Based Facilities

Two examples of ground-based facilities are the UKIRT and the 2MASS instruments. The UKIRT telescope[4] is a 3.8-m Cassegrain system. The UKIRT Wide Field Camera (WFCAM) has been designed for large-scale survey observations and uses cooled HgCdTe arrays. The *Two Micron All-Sky Survey* (2MASS)[5] has used two robotic telescopes with an aperture of 1.3 m and HgCdTe arrays to acquire photometric data in the J, H, and K bands for billions point sources. The survey has produced catalogs of point sources and extended sources. The operation of airborne and spaced instrumentation is affected by a background smaller by several orders of magnitude compared to the ground-based one. Aircrafts operating at high altitude are a cheaper

---

[3]http://www2.keck.hawaii.edu/inst/nirspec/.

[4]http://www.ukirt.hawaii.edu/.

[5]http://irsa.ipac.caltech.edu/Missions/2mass.html.

alternative to operation in space. The aircraft altitude offers a low temperature and a very low humidity. The *Stratospheric Observatory for Infrared Astronomy* (SOFIA)[6] is a Boeing aircraft with a Cassegrain telescope mounted in an open space close to the tail, while the instrumentation is in the pressurized section.

Infrared astronomy benefits more than optical astronomy from operation in space, since it provides access to regions not accessible to ground-based facilities. The whole instrumentation can be cooled at cryogenic temperature. The *InfraRed Astronomical Satellite* (IRAS)[7] has pioneered the infrared surveys in space, with its 0.6-m telescope and an array of extrinsic photoconductors in the focal plane. The mission performed observations at 12, 25, 60, and 100 μm, producing sky maps and a catalog of point sources. The *Spitzer* space telescope[8] has been launched in 2003 into a solar orbit. It uses a 0.85 m mirror equipped with a set of three instruments. The Multiband Imaging Photometer (MIPS) has three detectors arrays of observations at different wavelengths: a Si:As arrays for 24 μm and Ge:Ga arrays for 70 and 160 μm observations. The InfraRed Array Camera (IRAC) has InSb and Si:As arrays for the region at a few μm. The InfraRed Spectrometer (IRS) provides low-to-moderate resolution in the interval from 5 to 38 μm. The mission has adopted the strategy of cooling with the combination of liquid cryogens and radiative cooling, that has allowed the continuation of operation after the end of the cryogenic liquids. The *Herschel Space Telescope*[9] uses a telescope with an aperture of 3.5 m and is equipped with three instruments for the infrared. The Spectral and Photometric Imaging Receiver (SPIRE) comprises an imaging photometer and an imaging Fourier transform spectrometer; the detectors are bolometers with neutron transmutation-doped germanium thermometers. The Photodetector Array Camera and Spectrometer (PACS) is an imager (bolometer arrays) and spectrometer (Ge:Ga arrays) for the region between 60 and 210 μm. The Heterodyne Instrument for the Far Infrared (HIFI) is an heterodyne spectrometer that converts the frequency of the input signal to a lower value by the mixing with a reference local oscillator; this technique is widely used in radio astronomy (Chap. 9).

Problems

8.1 Discuss the different types of infrared detectors.

8.2 Discuss the operation of bolometer detectors.

8.3 Discuss the differences between the infrared arrays and the CCDs.

---

[6]https://www.sofia.usra.edu/.

[7]http://irsa.ipac.caltech.edu/IRASdocs/iras.html.

[8]http://www.spitzer.caltech.edu/.

[9]http://www.herschel.caltech.edu/.

8.4 Estimate the resolution power of the Hubble Space Telescope that has a primary mirror with a diameter of 2.4 m and is operating at the diffraction limit, at the near-infrared wavelength of 2.2 microns.

References

1. Glass, I. S., Handbook of Infrared Astronomy, Cambridge University Press (1999)
2. Lèna, P. et al.: Observational Astrophysics. Springer-Verlag Berlin Heidelberg (2012)
3. MCLean. I. S.: Electronic Imaging in Astronomy: Detectors and Instrumentation, Springer (2008)
4. Oswalt, T. D., McLean, I. S.: Planets, Stars and Stellar Systems. Volume I: Telescopes and Instrumentation. Springer (2013)
5. Oswalt, T. D., Bond, H. E.: Planets, Stars and Stellar Systems. Volume II: Astronomical Techniques, Software, and Data. Springer (2013)
6. Zombeck, M. V.: Handbook of Space Astronomy and Astrophysics. Cambridge, UK: Cambridge University Press (2007)

Chapter 9
Radio and Submillimeter Astronomy: Radio Telescopes

Radio astronomy, the first astronomy after the optical one, investigates the radio emission of celestial sources. This chapter discusses the receiver systems based on the superheterodyne technique, investigating the different components: antennas, amplifiers, mixers, and detectors. The faintness of the radio signals demands radio telescope systems with high amplification factors.

9.1 Astrophysical Radio Sources

Radio astronomy, investigating the radio emission of celestial sources, has been the first astronomy after the optical one. The radio window span a wide region of the electromagnetic spectrum, ranging from a wavelength of about 1 mm to a wavelength of about 15 m or, in units of frequency, from hundreds GHz down to a few MHz [5]. The low-frequency cutoff is due to plasma frequency cut-off in the ionosphere and depends on the epoch, the observatory location, and on the solar activity. The atmospheric transparency at the ALMA Observatory is shown in Fig. 9.1 for frequencies above 10 GHz. The transmission is high up to a few hundred Hz, with regions of absorption by water (22 and 183 GHz) and molecular oxygen (60 and 119 GHz).

The black body distribution in the radio region can be approximated by the Rayleigh–Jeans law. Thus, it is possible to define a *brightness temperature* T_B for a system with brightness B_ν:

$$T_B = \frac{c^2}{2\nu^2 k} B_\nu \tag{9.1}$$

The brightness temperature is a useful tool, even when the observed sources do not emit by thermal processes. The spectra of different sources are reported in Fig. 9.2. The brightness of spans a wide range, from mJy to 10^6 Jy. Since 1 Jy corresponds to 10^{-26} W m^{-2} Hz^{-1}, even with a large collecting area of the order of 100 m^2 and

© Springer International Publishing Switzerland 2017
R. Poggiani, *Optical, Infrared and Radio Astronomy*,
UNITEXT for Physics, DOI 10.1007/978-3-319-44732-2_9

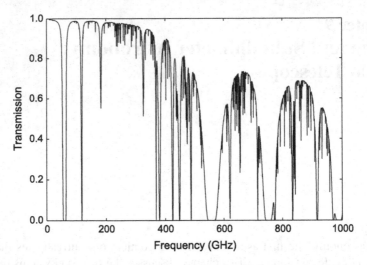

Fig. 9.1 Transparency of sky at the ALMA observatory (http://www.almaobservatory.org/)

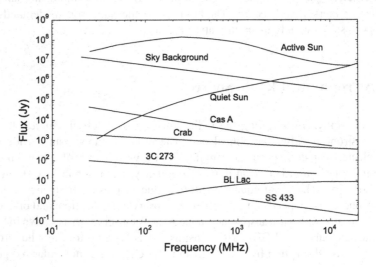

Fig. 9.2 Flux of radio sources (after [1])

a receiver bandwidth of about 1 MHz the received power of the order of 10^{-18} W. Thus, a receiver for radio astronomy must include, in addition to a large collecting area and a large bandwidth, also a large internal amplification factor.

The spectra can be described by a spectral index α, where the flux spectral density behaves as an inverse power law in frequency. The radio source can be divided into two families according to the value of the power:

- *thermal sources*: $\alpha \simeq -2$
- *non-thermal sources*: $\alpha \geq 0$

Some astronomical sources emit by the *synchrotron radiation* produced by high-energy electrons moving at relativistic speed in the interstellar magnetic field. The spectrum depends on the energy distribution of the electron population, but generally the brightness is proportional to the reciprocal of the frequency. Standard stars are weak radio emitters. since the black body flux behaves as the inverse square of the wavelength. Assuming a flux of the order of 10^4 Jy for the Sun (when not in the active phase), placing it at a distance of a few pc would drop the flux well below the μJy threshold.

The sources in the submillimeter astronomy include star formation regions and young galaxies. The submillimeter region is a bridge between infrared and radio astronomy, and the observational techniques are inspired by the instrumentation used in the two neighboring domains.

This chapter discusses the receiver systems based on the superheterodyne technique, investigating the different components: antennas, amplifiers, mixers, and detectors.

9.2 Radio and Submillimeter Receivers: Superheterodyne Detection

Radio astronomy [2–5] receivers detect the electric field of the incoming waves using an antenna as the first stage. Detection can be coherent or incoherent. *Coherent detection systems* preserve the information of the phase of the wave and are the only solution for interferometric observations. The standard radio astronomy receiver is the *superheterodyne system* (Fig. 9.3) that provides a first amplification of the input signal and its conversion to a lower frequency, the *Intermediate Frequency*. Further stages provide additional amplification and the detection.

The first stage is an *antenna*, a transducer to convert the electric field of the incident wave into an electric signal. The next stage is the *preamplification system* that must have high gains and low noise. The amplification stage is followed by the *mixer*, a device that combines the frequency of the incoming signal with the frequency of a *local oscillator*, to produce a signal at lower frequency, the intermediate frequency. At frequencies above 100 GHz, the mixer is the first stage after the antenna. The initial stages of the detection system operate at high frequency and form the *front end*. The intermediate frequency signal is fed to the *back end* that includes further amplification stages and the instrumentation to measure the time series or the spectrum of the signal.

The concept of an equivalent temperature is used to describe the noise performances of the receiver system. The power per unit bandwidth P_v produced by the Johnson noise in a resistor is proportional to the *noise temperature* T_N [5]:

$$P_v = kT_n \tag{9.2}$$

The limit of coherent receiver systems is set by assuming a power per unit band of $P_\nu = h\nu$; the minimum noise temperature is $T_N = \frac{h\nu}{kT}$. *Incoherent detection systems* provide a measurement of the total power, generally using bolometers, and cannot be used for interferometry. In the following, the components of the receiver system will be presented.

9.3   Antennas for Radioastronomy

The antenna is the first element of a radio astronomy receiver, to transform the electromagnetic field of the incoming wave into an electrical signal [2–5]. Antennas fulfill the *Reciprocity Theorem*: their properties are identical, for transmission and for reception. Antennas are *Hertz dipoles*; an element with a length l carrying a current I oscillating with a total wavelength λ will radiate a total power [5]:

$$P = \frac{2c}{3} \left(\frac{Il}{2\lambda} \right)^2 \tag{9.3}$$

The response of an antenna is defined by the *power pattern P*, the absolute value of the Poynting flux S. The power pattern of a dipole is shaped as a doughnut with a plane of symmetry orthogonal to the dipole. The directivity of the response is improved by the addition of a reflecting plane. The radiation of a dipole is linearly polarized, with the electric field directed along the dipole direction. A combination

Fig. 9.3 Superheterodyne receiver

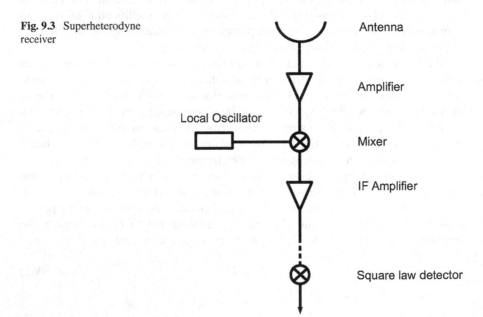

of two crossed dipoles is used to measure both polarization. The *beam solid angle* of an antenna is:

$$\Omega_b = \int P_n(\theta, \phi) d\Omega \qquad (9.4)$$

where $P_n(\theta, \phi)$ is the normalized power pattern. The *main beam* defines the solid angle containing the power pattern out to the first zero and defines the *main lobe* of the antenna:

$$\Omega_{mb} = \int_{main\ lobe} P_n(\theta, \phi) d\Omega \qquad (9.5)$$

The ratio of the main beam to the beam solid angle is the *beam efficiency* $\eta_b = \frac{\Omega_{mb}}{\Omega_b}$. The *antenna gain* is proportional to the power pattern:

$$G = \frac{4\pi P(\theta, \phi)}{\int P(\theta, \phi) d\Omega} \qquad (9.6)$$

For reference, the gain of a dipole is 1.76 dBi. The power pattern of a real antenna has a main lobe and some side lobes, in analogy to the central peak and the alternation of bright and dark rings of the diffraction pattern discussed for optical telescopes. The side lobes must be minimized to avoid the contamination of observations by the radiation of the surrounding environment and the ground. The collecting area of an antenna is defined by the *effective area* A_e, the ratio of the collected power to the absolute value of the Poynting flux:

$$A_e = G \frac{\lambda^2}{4\pi} \qquad (9.7)$$

The effective area is related to the geometrical area A_g by the *aperture efficiency*, according to $A_e = \eta_a A_g$. The collecting area is larger at large wavelengths, where simple dipoles or arrays of dipoles are used. For wavelengths smaller than about one meter, parabolic reflectors are used. The dishes of large radio telescopes are indeed reflectors: the real antenna is a dipole placed at the focus. The dipole is inside a waveguide ending into a *feed horn* that attenuates the effects of the side lobes. The waveguide is closed at the other end by a ground plane to reflect the unused lobe of the dipole. The waveguide allows a single mode of the radiation, acting as a high-pass filter. The steerable parabolic dishes used for radio astronomy have, to date, diameters up to 100 m. Due to the size and weight, radio dishes are pointed using alt-azimuth mountings. The largest radio telescope is the Arecibo reflector, with a diameter of 300 m, and not steerable. The constraints discussed in the optical domain about surface accuracy can be extended to the radio region. The level of tolerance on the optical surfaces, $\sim \frac{\lambda}{10}$, is easily achieved in the construction stage, given the large wavelengths of radio astronomy. The parabolic shape is realized assembling a combination of tunable panels. The desired shape must be maintained during pointing at any inclination. Thermal deformations are minimized by groups of warm air ducts and a careful selection of materials. The radio telescopes use the Cassegrain, the Gregorian, or Nasmyth configurations discussed in the optical domain, to achieve an easier access to instrumentation.

Antennas with and without reflectors can be discussed using the diffraction theory. The amplitude of the electric field produced by a current is the Fourier transform of the current distribution. An horn antenna or an array of dipoles can be modeled as a rectangular aperture with dimensions L_x and L_y, with a pattern showing a central bright region with dimensions $\frac{\lambda}{L_x}$ and $\frac{\lambda}{L_y}$. A parabolic reflector can be modeled as a circular aperture; thus, the power pattern will be an Airy pattern, showing a main lobe with a width of about $\frac{\lambda}{D}$ and *sidelobes* corresponding to the secondary maxima. The angular resolution of the system is worse in the radio than in the optical domain, due to the larger wavelengths of the radio domain, centimeters to meters, against the hundreds of nanometers of the optical domain. However, the use of different antennas combined in an interferometer with a very large baseline allows to achieve high angular resolution (Chaps. 11, 12).

The antenna response determines the response of a receiver. The power received from a source with a brightness distribution $B_\nu(\theta, \phi)$ and collected with an antenna with a power pattern $P_n(\theta, \phi)$ normalized to the maximum is:

$$P_\nu = \frac{1}{2} A_e \int B_\nu(\theta, \phi) P_n(\theta, \phi) d\Omega \tag{9.8}$$

where the factor $\frac{1}{2}$ accounts for the detection of a single polarization. The power per unit frequency from an antenna can be described by an equivalent *antenna temperature* T_a, the temperature of a matched resistor with a thermal power equal to the power from the antenna:

$$P_\nu = kT_a \tag{9.9}$$

The antenna temperature is a measurable quantity that is related to the convolution of the brightness temperature T_b of the observed object with the normalized power pattern (Fig. 9.4) [5]:

$$T_a(\theta^*, \phi^*) = \frac{\int T_b(\theta, \phi) P_n(\theta - \theta^*, \phi - \phi^*) \sin\theta \, d\theta \, d\phi}{\int P_n(\theta, \phi) d\Omega} \tag{9.10}$$

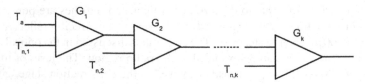

Fig. 9.4 Cascade of amplifiers

The brightness temperature is extracted by the inversion of the above equation. The power collected in the observation of sources much smaller than the beam is:

$$P_\nu = \frac{1}{2} A_e B_\nu = k T_a \qquad (9.11)$$

Thus, the antenna temperature is:

$$T_a = \Gamma B_\nu \qquad (9.12)$$

where Γ is the sensitivity of the telescope measured in K Jy^{-1}.

9.4 Telescopes for Submillimeter Astronomy

The collecting systems of submillimeter astronomy are similar to the telescopes used in the close infrared region. Generally, two-mirror reflecting telescopes are used with Cassegrain or Gregorian configurations with alt-azimuth mounts. The primary mirror is fast, to reduce the size (and the cost) of the instrument. The small focal ratios make astigmatism the dominant aberration. The primary mirrors are generally segmented.

9.5 Amplifiers

As discussed above, the amplification needed for radio astronomy observations is very high. The high gain factors are achieved by a cascade of k amplifiers, each having a gain G_i and a noise temperature $T_{n,i}$ [3–5].

The total gain is the product of the single gains:

$$G = \Pi_{i=1}^{k} G_i \qquad (9.13)$$

The total noise temperature $T_{n,tot}$ of the cascade is:

$$T_{n,tot} = T_{n,1} + \frac{1}{G_1} T_{n,2} + \cdots \frac{1}{G_1 G_2 \ldots G_{k-1}} T_{n,k} \qquad (9.14)$$

Since the first stage of amplification is the dominant contributor, it must have a very low-noise temperature, while the constraint is relaxed for the following stages since their noise is weighted by the product of the gains of the previous stages. The wide range of radio astronomical observations requires the use of different technologies in the different regions [3–5]. The standard choice for low frequencies are the *GaAS FET* amplifiers cooled at cryogenic temperatures. The *High Electron Mobility Transistors* (HEMTs) are a modification of the standard FETs. A layer of intrinsic GaAs is added to produce a two-dimensional gas of electrons that shows an higher mobility compared to the three-dimensional gas of the standard FETs. The noise temperature ranges from a few Kelvin at a some GHz to some tens Kelvin for frequencies of the order of one hundred GHz.

9.6   Mixers

Mixers are nonlinear elements that decrease the frequency of the input signal, by combining it with the reference signal of a local oscillator [5]. The local oscillators are built using different technologies: synthesizers in the region from centimeter to meter and Gunn oscillators above 100 GHz. The mixers are the mandatory solution for the front end of receivers operating above 100 GHz, since it is very difficult to build low-noise amplifiers at high frequency. Mixers are mounted just after the antenna. The output of a mixer is proportional to the square of the sum of the two inputs, the signal $V_s \sin(2\pi \nu_s t + \phi_s)$ from the antenna and the signal $V_l \sin(2\pi \nu_l t + \phi_l)$ from the local oscillator; for simplicity, we have considered both to be pure sinusoidal functions at the respective frequencies ν_s and ν_l. The output of the mixer will be include a term related to the difference of the frequencies ν_s and ν_l and additional terms that include a DC component, a term related to the sum of the two frequencies and terms with the second harmonics of the input signal and of the local oscillator:

$$U \propto (V_s \sin(2\pi \nu_s t + \phi_s) + V_l \sin(2\pi \nu_l t + \phi_l))^2 \propto V_s V_l \sin\left[2\pi(\nu_s - \nu_l)t + \left(\phi_s - \phi_l + \frac{\pi}{2}\right)\right] + ...$$
$$(9.15)$$

By using filtering, it is possible to select the component related to the difference of frequencies, called *intermediate frequency* $\nu_{IF} = \nu_s - \nu_l$, that preserves the information of the signal amplitude, since it is proportional to V_s.

The *SIS mixers* are superconductor–insulator–superconductor junctions that rely on the small value of the energy gap 2Δ in superconductors, of the order of meV, comparable with the energy of the detected photons, some hundreds GHz. The absorption of a photon produces the tunneling of an electron through the barrier of the insulator. When the SIS junction is operating without a local oscillator, there is a jump in the current–voltage curve at a bias $\frac{2\Delta}{e}$. The addition of the power of the local oscillator produces the appearance of a stair case pattern, promoting the conversion of photons to a lower energy, i.e., to a lower frequency (Fig. 9.5).

The *Hot Electron Bolometers* (HEB) are heterodyne devices base on thin superconducting films with a very low noise.

Amplifier and mixers are the ingredients of the front ends of the receiver systems. The HEMTs are used below 100 GHz, where they achieve noise temperatures of a few Kelvins. The region from 100 to 1000 GHz is the domain of SIS mixers cooled at cryogenic temperatures, whose noise temperature is of the order of tens to thousands Kelvins. The submillimeter region, above 1 THz, is covered by the superconducting

Hot Electron Bolometers (HEB), with noise temperatures of the order of thousands Kelvins.

9.7 Detectors

The heterodyne receivers detect the amplified and frequency-shifted signal using *square law detectors* [3–5]:

$$y(t) = ax^2(t) \tag{9.16}$$

whose output $y(t)$ proportional to the square of the input $x(t)$. An incoherent receiver system is a linear function of the detected power and acts as a *radiometer*. Since the expectation value of the output is proportional to the variance of the input, there will be a nonzero output even in absence of an input signal. The signals from antennas in radio interferometers are digitized to preserve the phase information before being sent to the correlator system in interferometers (Chaps. 11, 12).

9.8 Calibration

The heterodyne receivers are calibrated by producing a known temperature rise at the input and measuring the output [3–5]. The calibrators usually are two resistive loads at temperatures T_l and T_h. If the temperature of the receiver is T_r, the ratio of the two outputs out_h, out_l is the y factor:

$$y = \frac{out_h}{out_l} \tag{9.17}$$

Fig. 9.5 Curve of SIS mixers

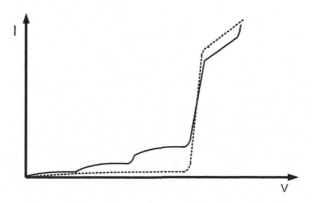

Thus, the receiver temperature is:

$$T_r = \frac{T_h - yT_l}{y - 1} \tag{9.18}$$

The resistive loads are kept at ambient temperature or at the temperature of liquid nitrogen or liquid helium. The method is used as a primary calibration method.

Problem

9.1 Estimate the minimum noise for coherent receivers operating at 100 GHz, 1 THz, 10^{14} Hz.

References

1. Allen, C. W. (Cox, A. N., Editor) : Allen's Astrophysical Quantities. Springer-Verlag (2013)
2. Lèna, P. et al.: Observational Astrophysics. Springer-Verlag Berlin Heidelberg (2012)
3. Oswalt, T. D., McLean, I. S.: Planets, Stars and Stellar Systems. Volume I: Telescopes and Instrumentation. Springer (2013)
4. Oswalt, T. D., Bond, H. E.: Planets, Stars and Stellar Systems. Volume II: Astronomical Techniques, Software, and Data. Springer (2013)
5. Wilson, T. L., Rohlfs, K. and Hüttemeister, S.: Tools of Radio Astronomy, Springer (2009)

Chapter 10
Radio and Submillimeter Astronomy: Receivers and Spectrometers

This chapter presents the different radio receiver configurations used to achieve the high stability demanded by the high gains required for the detection. The radio spectrometers are discussed. The chapter is closed by a discussion of the measurements of the CMB.

10.1 Total Power Radiometers

The simplest example of radio receiver is the *total power receiver* [1–4].

As discussed in the previous chapter, all noise contributions can be described by a suitable temperature. The receiver can be characterized by a *system temperature*, the sum of different effective temperatures T_i related to the emission from the source, the atmosphere, and the astrophysical background, to the noise of the receiver instrumentation and so on:

$$T_{sys} = \sum_i T_i \qquad (10.1)$$

The system temperature is the sum of the receiver temperature T_r and of the antenna temperature T_a:

$$T_{sys} = T_r + T_a \qquad (10.2)$$

For a receiver with a system temperature T_{sys}, the root-mean-square noise is the ratio of the mean value to the number of samples secured by a system with a bandwidth Δv during an interval τ:

$$\Delta T = \frac{T_{sys}}{\sqrt{\Delta v \tau}} \qquad (10.3)$$

© Springer International Publishing Switzerland 2017
R. Poggiani, *Optical, Infrared and Radio Astronomy*,
UNITEXT for Physics, DOI 10.1007/978-3-319-44732-2_10

Considering the detection of a single polarization, the rms variation of the flux density is [4]:

$$\Delta S = \frac{2kT_{sys}}{A_e\sqrt{\Delta\nu\tau}} \qquad (10.4)$$

The sensitivity depends on the collection area, on the bandwidth, and on the observation time.

10.2   Dicke Switching

The gain of a radio astronomy receiver should be stable, but the large gains used to amplify the weak signals make stability difficult. The relative fluctuation of the detected temperature is of the order of the relative fluctuation of the gain [4]:

$$\frac{\Delta T}{T_r + T_a} = \frac{\Delta G}{G} \qquad (10.5)$$

The *Dicke switch* technique measures alternatively the astronomical source, which produces an antenna temperature T_a, and a reference load, a resistor cooled with liquid nitrogen or liquid helium at a temperature T_c. The Dicke configuration allows to weight the gain fluctuation with a factor related to the difference between the antenna temperature and the reference load temperature:

$$\frac{\Delta T}{T_r} = \frac{\Delta G}{G}\frac{T_a - T_c}{T_r} \qquad (10.6)$$

Assuming that the source and the load are measured with identical times, the sensitivity is [4]:

$$\frac{\Delta T}{T_{sys}} = \frac{\sqrt{2}}{\sqrt{\Delta\nu\tau}} \qquad (10.7)$$

10.3   Correlation Receiver

The *correlation receiver* shown in Fig. 10.1 is an approach to improve the stability [4] that uses the properties of the *cross-correlation function* that describes the overlap of two functions $x(t)$, $y(t)$:

$$CCF(t) = \int x(t)y(t + \tau)d\tau \qquad (10.8)$$

The product of two functions is built using the algebraic operation, as $x \cdot y = (x + y)^2 - (x - y)^2$. The *3 dB hybrid* is a circuit element that accepts the inputs x and y and produces the outputs $\frac{x+y}{2}$ and $\frac{x-y}{2}$.

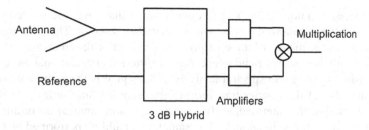

Fig. 10.1 Correlation receiver

The two signals in a correlation receiver are the signal from the antenna V_a and a reference signal V_c. The outputs at the exit ports of the 2 dB hybrid will be:

$$V_1 \propto \frac{V_a + V_c}{\sqrt{2}} + V_{N,1} \tag{10.9}$$

$$V_2 \propto \frac{V_a - V_c}{\sqrt{2}} + V_{N,2} \tag{10.10}$$

where $V_{N,i}$ are the noises in the two channels. Taking the correlation product of the two objects and performing a time average:

$$< V_1 V_2 > \propto < V_a >^2 - < V_c >^2 \tag{10.11}$$

since the noises are not correlated with the signals and between each other. The above term is the weight factor for the gain instability.

The sensitivity is identical to that of the Dicke switch [4]:

$$\frac{\Delta T}{T_{sys}} = \frac{\sqrt{2}}{\sqrt{\Delta \nu \tau}} \tag{10.12}$$

The correlation receiver is used to measure the polarization. The two orthogonal polarizations are fed to the inputs of a correlation receiver. The correlation of the signals without and with a phase delay in one of the channel is proportional to the U, V Stokes parameters.

10.4 Radio Spectrometers

Spectroscopy with superheterodyne receivers [1–4] can be performed by adding a scanning filter after the intermediate frequency stage. The filter scans the desired spectral band by tuning the center frequency of a narrowband filter. The solution is relatively cheap, since a single filter and a single square law detector are used. An extension of the technique is the use of *filter banks*, a set of adjacent filters with closely spaced and not overlapping frequency bands that cover the desired spectral band. Each filter is followed by a separate square law detector. The combination of channels produces the distribution of power a function of the frequency. Different filter banks are available at radio telescopes for different observational modes.

A possible approach to spectral analysis of the signal is the computation of the square modulus of the Fourier transform of the recorded time series [4]. The signal is extracted at the intermediate frequency stage and sampled according to the Nyquist theorem: for a signal at 5 GHz, sampling should be performed at 10 GHz. The approach has the advantage of allowing the measure of the signal power without the need of an external calibration system.

The sensitivity of a spectrometer is given by [4]:

$$\frac{\Delta T}{T_{sys}} = \frac{\sqrt{k}}{\sqrt{\Delta v \tau}} \tag{10.13}$$

where k is a factor that summarizes the specific properties of the instrumentation. Since the bandwidth of spectroscopic observations is much smaller than the bandwidth of observations of total power, comparable performances are achieved only by using long integration times.

10.4.1 Autocorrelation Spectrometers

Fourier transform is the basis of the *autocorrelation spectrometer* [1–4]. According to the Wiener–Kinchine Theorem, the power spectrum $S(v)$ is the Fourier transform of the autocorrelation function $R(\tau)$ of the signal $x(t)$:

$$R(\tau) = \int x(t + \tau) x(t) dt \tag{10.14}$$

$$s(v) = \int R(\tau) \exp(-2\pi i v t) d\tau \tag{10.15}$$

The signal $x(t)$ is sampled and then transformed to a digital signal $w(x)$ using a *clipper* to provide one-bit quantization. The output signal has two levels only: 1 for the positive part and 0 elsewhere. The autocorrelation functions $R_x(\tau)$ and $R_w(\tau)$ of the input signal $x(t)$ and the clipped signal $w(t)$ are related by:

$$R_x(\tau) = R_x(0) \sin\left[\frac{\pi}{2} R_w(\tau)\right] \tag{10.16}$$

The measure of the autocorrelation function of the clipped signal $w(t)$ provides a measure of the autocorrelation function of the input signal $x(t)$. The hardware implementation is shown in Fig. 10.2. The clipped samples are sent to a group of shift register providing a time shift of $\Delta \tau$ each. The original unshifted signal is

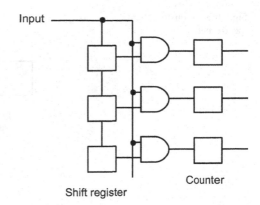

Fig. 10.2 Autocorrelation spectrometer

compared with the time-shifted version using an AND logical element. The output of the different coincidences is proportional to the autocorrelation function $R_w(\tau)$.

The amplitude information of the signal is lost in the clipping process, and a calibration of the total power must be provided by an independent radiometer. The autocorrelation spectrometer has found a widespread application, due to the easy implementation from the point of view of electronics. The resolution of the autocorrelation spectrometer is [4]:

$$\frac{\Delta T}{T_{sys}} = \frac{\frac{\pi}{2}}{\sqrt{\Delta\nu\tau}} \tag{10.17}$$

10.4.2   Acousto-Optical Spectrometers

The *acousto-optical spectrometers* (Fig. 10.3) use the diffraction of radiation on ultrasonic waves [4]. Acoustic waves produce periodic variations of the density and of the index of refraction of a medium that will act as a three-imensional grating with a spacing Λ. The grating will appear as stationary, due to the low compression velocity, to an incident wave, that will be diffracted. The incidence angle α_i and the diffraction angle α_d of radiation with a wavelength λ fulfill a Bragg condition:

$$\Lambda(\sin\alpha_d - \sin\alpha_i) = \lambda l \tag{10.18}$$

where l is an integer number. The signal from the receiver is used to drive a Bragg cell to produce a periodic structure. The radiation of a laser is sent to the cell. The intensity of the diffracted wave, measured with a CCD, is proportional to the power spectrum of the radiation.

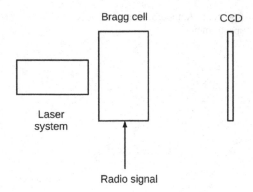

Fig. 10.3 Acousto-optical spectrometer

10.5   Large Facilities

The large radio telescopes are equipped with a large number of receivers and spectrometers for different regions of the spectrum whose characteristics are available at the observatory sites. Some large radio telescopes are mentioned as examples. The *Parkes radio telescope*[1] uses a 64-m parabolic dish. Several feeds and receivers are available, from 700 MHz to 15 GHz; among them, the multibeam receiver for the 21-cm radiation that has been extensively used for survey and pulsar searches. The *Effelsberg radio telescope*[2] uses a large steerable parabolic dish, with a diameter of 100 m, and a secondary with a size of 6.5 m. The frequency coverage is very wide, from hundreds MHz to about 100 GHz. The instrument has been used for mapping the water vapor line at 22 GHz. Together with the Parkes radio telescope, the Effelsberg telescope has produced a detailed map of the radio emission at 408 MHz. The *Arecibo Observatory*[3] hosts the largest dish ever built, with a diameter of 300 m. The dish is a fixed spherical reflector that is fixed, but its focus can be moved. The available receivers cover the region from hundreds MHz to some GHz.

10.6   CMB Observations

The observations of the Cosmic Microwave Background (CMB) involve the measurement of the absolute temperature and of its fluctuations and are performed with ground-based, balloon born, or space-based instruments [2, 3]. The CMB is investigated in the frequency region from tens to hundreds GHz, where the noises from the synchrotron emission and the dust are smaller.

---

[1] https://www.parkes.atnf.csiro.au/.

[2] http://www.mpifr-bonn.mpg.de/en/effelsberg.

[3] https://www.naic.edu/ch.

The measurement of the CMB fluctuations deals with effects as small a few parts in 10^7. The angular resolution of the CMB instrumentation has steadily improved from the initial 7^0 for COBE, being now in the range of a few arc minutes for WMAP, Planck, and ACT. The instrumentation combines several features of the optical and radio astronomies. Most experiment for CMB observation adopt reflective optics, a combination of reflectors with Gregorian configurations (see, e.g., the MAXIMA experiment).[4] The most notable exception is BICEP2[5] that has adopted refractive optics; the detection system of the experiment, operating at 150 GHz, is composed of bolometers sensitive to polarization, made of transition edge sensors coupled to phased antenna arrays. The satellites for the observations of CMB anisotropy have produced all sky maps of the sky temperature. The *Cosmic Background Explorer* (COBE) satellite[6] had three instruments on board: the Differential Microwave Radiometer (DMR), the Far Infrared Absolute Spectrophotometer (FIRAS), and the Diffuse Infrared Background Experiment (DIRBE). The FIRAS instrument has measured the absolute temperature of the CMB. The DMR instrument was composed of six receivers, each consisting of two feed horns (with separate calibrators) that sent the signal to a superheterodyne receiver operating with a Dicke switch configuration. The *Wilkinson Microwave Anisotropy Probe* (WMAP)[7] has two reflectors with a dimension of 1.6×1.4 m and an axis separation of 141^0 to perform differential measurements. The reflectors focus the radiation onto ten feeds. The *Planck* mission[8] has an aplanatic Gregorian configuration and hosts two different instruments to cover a wide part of the radio and millimeter part of the spectrum. The low-frequency region (30 to 70 GHz) is observed with cooled coherent detectors and a differential receiver with a cold load on one input; the high-frequency region (90 to 900 GHz) is observed with bolometers operated at milliKelvin temperatures in a Fourier transform spectrometer configuration. The measurement of the polarization of the CMB is related to the early universe history. The effect is much smaller than the temperature anisotropies and is very sensitive to the contamination from the background. The instrumentation for measuring the polarization in Planck and WMAP is based on correlation receivers, where the polarization is estimated by the difference in the power related to the two polarizations.

Problem

10.1 A radio astronomy receiver has a bandwidth of 50 MHz. Estimate the integration time needed to achieve a rms noise that is 1 % of the total noise with a total power radiometer.

---

[4]http://cosmology.berkeley.edu/group/cmb/.

[5]https://www.cfa.harvard.edu/CMB/bicep2/.

[6]http://lambda.gsfc.nasa.gov/product/cobe/.

[7]http://map.gsfc.nasa.gov/.

[8]http://www.cosmos.esa.int/web/planck.

References

1. Lèna, P. et al.: Observational Astrophysics. Springer-Verlag Berlin Heidelberg (2012)
2. Oswalt, T. D., McLean, I. S.: Planets, Stars and Stellar Systems. Volume I: Telescopes and Instrumentation. Springer (2013)
3. Oswalt, T. D., Bond, H. E.: Planets, Stars and Stellar Systems. Volume II: Astronomical Techniques, Software, and Data. Springer (2013)
4. Wilson, T. L., Rohlfs, K. and Hüttemeister, S.: Tools of Radio Astronomy, Springer (2009)

Part IV
Instruments Acting Together: Interferometry

Chapter 11
Interferometry and Aperture Synthesis

Interferometry with arrays of telescopes or antennas achieves angular resolutions that are not in the reach of single instruments. The present chapter discusses the principles of interferometry that are the foundations of both optical/infrared and radio interferometry. We will firstly discuss the concept of coherence and the van Cittert–Zernike theorem that gives the fundamental equation for interferometry. The technique of aperture synthesis is discussed, together with the methods of image reconstruction.

11.1 Coherence

The radiation from astronomical sources is not totally coherent [4–6, 8]. The correlations in time and in space of the electric field are described by the *coherence function*, the cross-product of the field at the position x_1 at time t_1 and at x_2 at t_2:

$$V(r_1, t_1, r_2, t_2) = < E(r_1, t_1) \times E^*(r_2, t_2) > \tag{11.1}$$

where the average is performed over a time interval longer than the wave period. The coherence function can be discussed by firstly considering the *spatial coherence function*, where $t_1 = t_2$, and the *temporal coherence function*, where $r_1 = r_2$. The spatial coherence function measures the correlation between the field values at different positions and depends on a single parameter, the separation $\rho = r_1 - r_2$ of the points r_1, r_2:

$$V(r_1, t_1, r_2, t_1) = V(r_1 - r_2) = V(\rho) \tag{11.2}$$

Interference fringes are observed with non-monochromatic radiation with a bandwidth $\Delta \nu$ if the difference of the optical paths is smaller than the *coherence length* $l_c = \frac{c}{\Delta \nu}$.

© Springer International Publishing Switzerland 2017
R. Poggiani, *Optical, Infrared and Radio Astronomy*,
UNITEXT for Physics, DOI 10.1007/978-3-319-44732-2_11

The temporal coherence function describes the correlation of the field at different epochs and depends on a single parameter, the difference $\tau = t_1 - t_2$ of times:

$$V(r_1, t_1, r_1, t_2) = V(t_1 - t_2) = V(\tau) \tag{11.3}$$

The timescale of the temporal coherence function is the *coherence time* $\tau_c \sim \frac{1}{\Delta\nu}$, where $\Delta\nu$ is the bandwidth of the spectral distribution of the radiation.

An interferometer is an instrument that measures the spatial coherence function. The spatial intensity distribution of the radiation from an astronomical source is reconstructed by the measurements of the spatial coherence function measured at different points.

11.2   The van Cittert–Zernike Theorem

The *van Cittert–Zernike Theorem* [5, 6, 8] defines the *complex visibility* $\hat{V}$ that can be built from the intensity distribution $I(\sigma)$ on the sky using two apertures (telescopes and antennas) separated by the *baseline* **B**:

$$V = \int A(\sigma)I(\sigma)e^{-\frac{2\pi i}{\lambda}\mathbf{B}\cdot\sigma}d\Omega \tag{11.4}$$

where σ is the vector from the center of the observed field to a point in the sky, and $A(\sigma)$ is the normalized response function of the antenna or the telescope. The baseline and the σ vector are measured using equatorial East–North coordinates. The modulus of the complex visibility function is called *fringe amplitude* or *visibility*. Defining the components of the baseline and of the vector σ in units of the wavelength of observation as $\mathbf{u} = (u, v)$ and $\sigma = (l, m)$, the visibility functions become:

$$\hat{V}(u, v) = \int_{l,m} A(l, m)I(l, m)e^{-2\pi i(ul+vm)}dldm \tag{11.5}$$

An interferometer probes an intensity distribution by measuring the Fourier components $\hat{V}(u, v)$ for a large number of baselines. Due to the properties of the Fourier transform, the components allow the reconstruction of the original image by an inverse Fourier transform. The simplest interferometer has two elements and a single defining baseline B; thus, it will probe a spatial frequency $\frac{B}{\lambda}$. The angular resolution of an interferometer is given by $\frac{\lambda}{B_{max}}$, where B_{max} is the length of the largest baseline.Interferometers operating with visible light at about 500 nm can achieve

resolution of 10 or 1 mas with baselines of the order of 10 and 100 m. The radiation
from relatively small apertures positioned at large distances can provide an angular
resolution that can only be achieved by very large apertures.

11.3   Michelson Stellar Interferometer

The *Michelson stellar interferometer* is an example of interferometer that can be
modeled by a pair of apertures in an opaque screen that play the role of the pair of
telescopes in a real interferometer [4–6, 8]. If the system observes the radiation of
a binary source with an angular separation θ in the sky, the observed pattern will be
the combination of two patterns, each one being an interference pattern modulated
by the diffraction pattern of a single aperture. The resulting signal depends on the
separation of the sources. For a disk of with an angular size H the fringe pattern
disappears when:

$$d_a = 1.22 \frac{\lambda}{H} \tag{11.6}$$

where d_a is the separation of the apertures. The practical realization of the stellar
interferometer is shown in Fig. 11.1. The separation of the apertures is varied until
the fringes disappear. Today, the apertures are replaced by separate instruments.

Fig. 11.1 Michelson stellar
interferometer. The beams
from the mirrors M1, M2 are
tuned to produce the
superposition at the focus of
the telescope

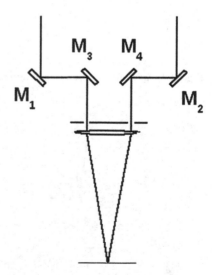

11.4   Aperture Synthesis

The technique of *aperture synthesis* [4–8] combines the information of pair of instruments with different baselines. According to the van Cittert–Zernike theorem, the signal of the pair measures one component of the Fourier transform of the observed field. The sky map is reconstructed using a large number of baseline combinations. Due to the Earth rotation, the orientation and the projection of the baseline change over 24 hours. The projected separation of two instruments traces an ellipse in the uv plane during 24 h; only 12 h are effectively required, using the complex conjugate of the data for the missing part. Different baselines track sets of concentric ellipses

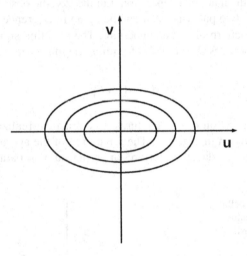

Fig. 11.2   Aperture synthesis: path of different baselines in the *uv* plane

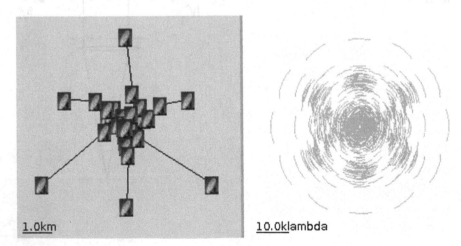

Fig. 11.3   The ASKAP array (*left*) and its uv coverage (*right*); data from the Virtual Radio Interferometer

(Fig. 11.2). The use of different baselines at the same time greatly shortens the time required to secure different Fourier components.

Examples of uv coverage for some interferometers can be obtained using the Virtual Radio Interferometer.[1] The uv coverage for the ASKAP instrument, an array of 36 antennas with a diameter of 12 m (Fig. 11.3, left) is shown in Fig. 11.3, right.

11.5 Image Reconstruction

Interferometry is used to reconstruct the morphology of astronomical sources [5, 6, 8]. The normalized brightness distribution is extracted by the inversion of Eq. 11.5:

$$I(l, m) = \int V(u, v) \exp^{+2\pi i(ul+vm)} du dv \qquad (11.7)$$

The practical application of the equation is difficult, since the visibility function $V(u, v)$ is obtained by sampling for a finite number of u, v points that can be summarized by a *sampling function* or *dirty beam* $C(u, v)$. The reconstructed brightness distribution is called *dirty map* and is the convolution of the Fourier transform of the dirty beam and of the real brightness distribution:

$$I_{dirty}(l, m) = \int C(u, v) V(u, v) e^{+2\pi i(ul+vm)} du dv = C_{dirty}(l, m) * S(l, m) \qquad (11.8)$$

The dirty beam acts as a Point Spread Function with the same role of the Airy pattern of a circular aperture, even if it is generally more complicate since it is related to the specific sampling strategy. The dirty map contains the effects of the side lobes of the Point Spread Function. The *deconvolution* of artifacts is traditionally performed with the *CLEAN* algorithm [3]. The dirty beam is normalized to the point of maximum intensity in the dirty map and subtracted from the latter. The process is iterated until the maximum intensity in the dirty map is comparable with the level of noise. The clean map is obtained by inserting all elements removed in the iterative subtraction process as *clean beams*, peaked functions with widths comparable with the response of the dirty beam. The algorithm is very effective for fields consisting of point-like sources.

The analysis of extended sources is performed with the *Maximum Entropy Method* (MEM) [1]. The algorithm requires that intensities cannot be negative and selects the smoothest solution consistent with data, maximizing the quantity:

$$S = -\sum p_i \ln p_i \qquad (11.9)$$

where $p_i = \frac{x_i}{\sum x_i}$, where x_i is the value to be analyzed. The algorithm selects the real features in the data, but it can be too selective and reject genuine ones.

---

[1] http://www.atnf.csiro.au/vlbi/calculator/.

11.6 Intensity Interferometry

The *intensity interferometer* [9] has been initially proposed in the context of radio astronomy, but has found its application in optical interferometry for the measurement of stellar diameters. The Narrabri interferometer in Namibia used two reflectors with an aperture of 6.5 m that could be moved along a circular track with a radius of 94 m, leading to a maximum baseline of 196 m. The intensity interferometer is based on the phase difference of the beating signals of distinct sources at the same instrument. The Fourier component at frequency f_1 of the signal from a point of the source is combined with the Fourier component at frequency f_2 of another point on the source at each instrument. The intensity components corresponding to the low beat frequency $f_1 - f_2$ that usually lie in the radio domain for optical interferometers are selected by filtering, and then multiplied and integrated in time to estimate the *correlation function* Fig. 11.4 shows the correlation function of an intensity interferometer and the visibility function of a Michelson interferometer for a disk of angular size H:

The correlation function has its first zero when the star has an angular diameter:

$$H = 1.22 \frac{\lambda}{d_j} \tag{11.10}$$

where d_j is the separation of the two instruments. The angular resolution is the same of a large telescope with a diameter equal to the separation of the telescopes.

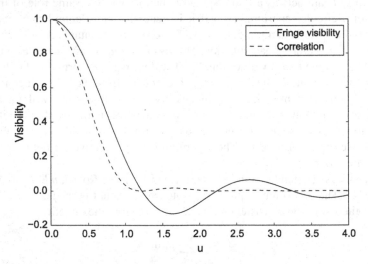

Fig. 11.4 Comparison of the fringe visibility observed in a Michelson interferometer with the correlation function measured by an intensity interferometer for a disk with diameter H; the quantity u is the ratio of the product of the disk size and the instrument separation to the radiation wavelength

References

1. Gull, S. F., Daniell, G. J.: Image reconstruction from incomplete and noisy data. Nat **272**, 686 (1978)
2. Hanbury-Brown, R., Twiss, R. Q.: Interferometry of Intensity Fluctuation in Light, Part I. Proc. R. Soc. A **242**, 300 (1957)
3. Hogbom, J. A.: Aperture Synthesis with a Non-Regular Distribution of Interferometer Baselines. A&ASS **15**, 417 (1974)
4. Lèna, P. et al.: Observational Astrophysics. Springer-Verlag Berlin Heidelberg (2012)
5. Oswalt, T. D., McLean, I. S.: Planets, Stars and Stellar Systems. Volume I: Telescopes and Instrumentation. Springer (2013)
6. Oswalt, T. D., Bond, H. E.: Planets, Stars and Stellar Systems. Volume II: Astronomical Techniques, Software, and Data. Springer (2013)
7. Ryle, M., Hewish, A.: The synthesis of large radio telescopes. MNRAS **120**, 220 (1960)
8. Thomson, A. R., Moran, J. M. and Swenson Jr, G.: Interferometry and Synthesis in Radio Astronomy. Wiley-VCH Verlag GmbH & Co. KGaA, Weinheim (2004)
9. Walker, G.: Astronomical Observations: an Optical Perspective. Cambridge University Press (1997)

Chapter 12
Interferometers

This chapter presents the technical and practical aspects of optical and radio interferometry. The two regions of the spectrum require different approaches for the processing and the combination of the signals from the single elements.

12.1 Optical and Infrared Interferometers

The working principle of optical and infrared interferometers is shown in Fig. 12.1 [1–4]; for clarity, a single baseline interferometer is considered. If the baseline is B, there is a geometric delay $B \sin \theta$ between the two telescopes when they are pointing to a source at a zenith angle θ. The delay is varying in time, due to the motion of the Earth. The light from the two apertures is sent to an optical path length equalizer, a delay that cancels the difference of the two optical paths, since it must be smaller than the coherence length of the radiation. The telescopes are usually equipped with adaptive optics systems or tip–tilt correctors at least to reduce the degradation due to the presence of the atmosphere. The sites of interferometers are chosen following the same prescriptions outlined for optical and infrared telescopes. The interferometer components must be shielded from external vibrations.

The optical radiation collected cannot be coherently amplified without a degradation of the signal-to-noise ratio [3, 4]. In the radio astronomy regime, the number of photons in each electromagnetic photons is very high and coherent amplification is feasible. According to the uncertainty principle, the relation between the number of photons and the phase is $\Delta \phi \Delta n \geq \frac{1}{2}$. A coherent amplifier adds a power noise whose minimum is:

$$P_m = h\nu + \frac{h\nu}{e^{\frac{h\nu}{k_b T}} - 1} \tag{12.1}$$

© Springer International Publishing Switzerland 2017
R. Poggiani, *Optical, Infrared and Radio Astronomy*,
UNITEXT for Physics, DOI 10.1007/978-3-319-44732-2_12

Fig. 12.1 Layout of an
optical/infrared
interferometer with two
elements

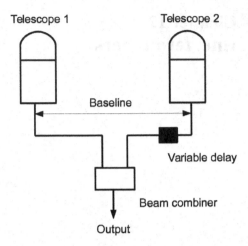

Radio astronomy and optical astronomy correspond to two different regimes. In the first one, high gains can be obtained by operating the amplifiers at low temperatures. In the optical regime, the noise corresponds to an equivalent temperatures of the order of thousands Kelvins.

The atmosphere is the dominant noise source for optical and infrared interferometers. The outer scale length L of the turbulence discussed in Chap. 4 is of the order of tens to hundreds meters. The effect of the atmosphere increases if the baseline is smaller than the outer scale length and becomes stable afterward.

Any source of noise will decrease the amplitude of the visibility function and will produce random variation of its phase. The phase error at each telescope depends on the local conditions of the atmosphere, but the visibility function is related to the differences of physical quantities at each site [3, 4]. If the phase at each instrument is φ_i, the phase of the visibility function between pairs of telescopes will be $\varphi_{ij} = \varphi_i - \varphi_j$. The atmosphere will contribute with an additional error ξ_i at each telescope. The phase difference between each pair of telescopes will be:

$$\delta_{ij} = \varphi_{ij} + \xi_i - \xi_j \tag{12.2}$$

In typical conditions, the errors produced by the atmosphere are larger than the intrinsic phase of the visibility function. The *closure phase* is the sum of the phases of a closed triangle of baselines:

$$\delta_{123} = \varphi_{12} + \varphi_{23} + \varphi_{31} \tag{12.3}$$

where the errors caused by the atmosphere cancel out. The closure phase is a good estimator of the visibility phase that is crucial to extract to reconstruct the brightness image by Fourier inversion. Usually, the mean closure phase is computed from the *triple correlation function* or bispectrum:

$$V_{123} = V_{12}V_{23}V_{31} = |V_{12}||V_{23}||V_{31}| \exp[i(\varphi_{12} + \varphi_{23} + \varphi_{31})] \tag{12.4}$$

Two relevant examples of interferometers are CHARA and VLTI. The *CHARA* array at Mount Wilson[1] is made of 6 telescopes with an aperture of 1 m and a range of baselines up to 330 m. The interferometer has a resolution of 200 μ as; it is extensively used to measure the stellar sizes.

The *Very Large Telescope Interferometer* (VLTI)[2] is composed of four 8.2-m Unit Telescopes (UT) and four 1.8-m Auxiliary Telescopes (AT). The UT instruments are at fixed locations, while the AT instruments can be positioned in thirty different locations. The combination of radiation can be performed for three or four telescopes.

12.2 Radio Interferometers

The simplest radio interferometer is made of two antennas (Fig. 12.2) [1, 3–5].

The incident wave with amplitude E produces the signals:

$$V_1 \propto E e^{i\omega t} \tag{12.5}$$

$$V_2 \propto E e^{i\omega(t-\tau)} \tag{12.6}$$

at the two instruments, where τ is the geometrical delay due to the angle between the baseline and the direction of the wave. After the amplification, the signals from each antenna are sent to a central unit where they are firstly corrected for the geometric delay caused by the Earth motion. The output of the two antennas is correlated and then integrated for a time T:

$$O(\tau) \propto E^2 \int_0^T e^{i\omega t} e^{-i\omega(t-\tau)} dt \propto E^2 e^{i\omega\tau} \tag{12.7}$$

The output of the interferometer is a periodic function of the delay τ that changes due to the motion of the Earth. The delay is the difference of the geometrical delay defined by the baseline vector **B** and the instrumental delay τ_i, according to: $\tau = \frac{1}{c}\mathbf{B}\cdot\mathbf{s} - \tau_i$, where **s** is the direction of observation. For a general brightness distribution I_ν measured with antennas with effective area A, the *visibility function* is given by a two-dimensional Fourier transform:

$$R(\mathbf{B}) = \int \int A(\mathbf{s}) I_\nu(\mathbf{s}) e^{i\omega(\frac{1}{c}\mathbf{B}\cdot\mathbf{s} - \tau_i)} d\Omega d\nu \tag{12.8}$$

---

[1]http://www.chara.gsu.edu/.

[2]https://www.eso.org/sci/facilities/paranal/telescopes/vlti.html.

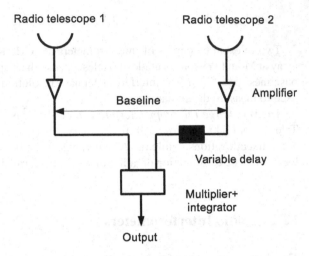

Fig. 12.2 Layout of radio interferometers

In an array with n antennas, there will be $N = \frac{n(n-1)}{2}$ simultaneous pairings of instruments. In analogy to the derivation of the sensitivity of single radio telescope reported in Eq. 10.3, the rms noise of the brightness temperature will be [3–5]:

$$\Delta T = \frac{2N k_B \lambda^2 T_{sys}}{A_e \Omega^* \sqrt{2N \Delta v \tau}} \tag{12.9}$$

where Ω^* is the dimension of the antenna beam. The rms variation of the flux density for an interferometer is:

$$\Delta S = \frac{2N k_B T_{sys}}{A_e \sqrt{2N \Delta v \tau}} \tag{12.10}$$

To achieve a high sensitivity, the elements of the interferometric array must have a large collection area. Usually, the parabolic reflectors have diameters in the range 20–30 m. The interferometers discussed so far use close antennas that are physically connected apertures. The *Very Long Baseline Interferometry* (VLBI) uses antennas at large distances; thus, they operate independently. The signals were originally stored on magnetic tapes using independent atomic clocks as the time reference. The data were then correlated at a single processing center. The evolution of the techniques has replaced the magnetic tapes with fiber optics links.

Several large interferometers are in operation or at the design stage. The *Very Large Array* (VLA)[3] in New Mexico is an interferometer with 27 Cassegrain antennas with a diameter of 25 m arranged as a Y-shaped array with a variety of baselines up to 36 km. The *Merlin*[4] system at Jodrell Bank is made of 7 antennas with a maximum baseline of 230 km.

---

[3]http://www.vla.nrao.edu/.

[4]http://www.e-merlin.ac.uk/.

The *Australia Telescope Compact Array* (ATCA)[5] is a system is with 6 dishes with a diameter of 22 m and baselines from 200 m to 6 km. The instrument operates with a group of receivers at different frequencies at the same time.

The *Atacama Large Millimeter/submillimeter Array* (ALMA),[6] located at the Chilean Andes at an elevation of about 5 km, is an array of fifty antennas with a diameter of 12 m, complemented by a dense array of twelve 7-m antennas and four 12-m antennas for the measurement of the total power. The instrumentation provides imaging and spectroscopy capability in the millimeter and sublimate domain.

The *Very Long Baseline Array* (VLBA)[7] is an array of ten dishes with a diameter of 25 m with baselines that are as long as over 8000 km from Hawaii to the Virgin Islands.

The *Square Kilometer Array* (SKA) is a future facility with a global collection area of about 1 km^2 with a frequency coverage from about 100 Hz to about 25 GHz with baselines up to 3000 km. The precursor is ASKAP.[8]

References

1. Lèna, P. et al.: Observational Astrophysics. Springer-Verlag Berlin Heidelberg (2012)
2. Monnier, J. D.: Optical Interferometry in Astronomy. RPP **66**, 789 (2003)
3. Oswalt, T. D., McLean, I. S.: Planets, Stars and Stellar Systems. Volume I: Telescopes and Instrumentation. Springer (2013)
4. Oswalt, T. D., Bond, H. E.: Planets, Stars and Stellar Systems. Volume II: Astronomical Techniques, Software, and Data. Springer (2013)
5. Thomson, A. R., Moran, J. M. and Swenson Jr, G.: Interferometry and Synthesis in Radio Astronomy. Wiley-VCH Verlag GmbH & Co. KGaA, Weinheim (2004)

---

[5] https://www.narrabri.atnf.csiro.au/.

[6] http://www.almaobservatory.org/.

[7] https://science.nrao.edu/facilities/vlba.

[8] http://www.atnf.csiro.au/projects/askap/index.html.

Part V
Observing

Chapter 13
Observations: Preparation and Execution

This chapter discusses the planning of observations and the choice of a telescope and of the epoch of observation. The signal-to-noise ratio for different types of observations is presented is discussed to estimate the exposure time. This chapter discusses the writing of observational proposals.

13.1   Target Selection: Coordinates and Finding Charts

The known astronomical objects can be searched in Simbad to access the right ascension and declination. Before starting observations, it is necessary to build the *finding charts*, maps of limited regions of the sky centered around a specific position with a size of a few arc minutes, typical of medium and large optical telescopes. An example is reported in Fig. 13.1.

13.2   Observations: Site and Epoch Selection

The sidereal time at midnight governs which celestial objects are transiting at the meridian at midnight and defines the range of accessible right ascensions at a chosen epoch. At the meridian, the sidereal time coincides with the right ascension of the objects. The objects should be observed at the meridian or close to it, when they will be at the maximum elevation above the horizon. The typical observing slot is an hour angle ranging from -3 to $+3$ h. For example, an object with a right ascension of 19 h will transit at the meridian on July 21 at 23 h.

© Springer International Publishing Switzerland 2017
R. Poggiani, *Optical, Infrared and Radio Astronomy*,
UNITEXT for Physics, DOI 10.1007/978-3-319-44732-2_13

Basic data :

3C 454.3 -- Quasar

Other object types: Rad (<u>Ref</u>,3C,...), QSO (<u>Ref</u>,QSO,...), gam (1AGL,1AGLR,...), X (<u>Ref</u>,1E,...), Bla
                       (<u>Ref</u>,[DGT2001],...), IR (2MASS,2MASSI,...), AGN (<u>Ref</u>,CGRaBS), smm (JCMTSE,JCMTSF), V*
                       (AAVSO), AG? (<u>Ref</u>), G (LEDA), UV (KUV)

ICRS coord. *(ep=J2000)* :   22 53 57.74798 +16 08 53.5611 (Radio) [0.11 0.04 0] A <u>2010AJ....139.1695L</u>
FK5 coord. *(ep=J2000 eq=2000)* : 22 53 57.748 +16 08 53.56 [0.11 0.04 0]
FK4 coord. *(ep=B1950 eq=1950)* :22 51 29.53 +15 52 54.2 [0.11 0.04 0]
Gal coord. *(ep=J2000)* :   086.1111 -38.1838 [0.11 0.04 0]
Radial velocity / Redshift / cz : V(km/s) 165232 [51] / z(~) 0.859001 [0.000170] / cz 257522.02 [50.96]
                                   D <u>2002LEDA.........0P</u>

Angular size *(arcmin)*:   0.0000122 0.0000052 8 (Rad) D <u>2010AJ....139.1713C</u>

Fluxes (6) :   B 16.57 [~] D <u>2010A&A...518A..10V</u>
                                 V 16.10 [~] D <u>2010A&A...518A..10V</u>
                                 R 15.22 [~] D <u>2008ApJS..175...97H</u>
                                 J 14.494 [0.030] C <u>2003yCat.2246....0C</u>
                                 H 13.855 [0.030] C <u>2003yCat.2246....0C</u>
                                 K 13.061 [0.027] C <u>2003yCat.2246....0C</u>

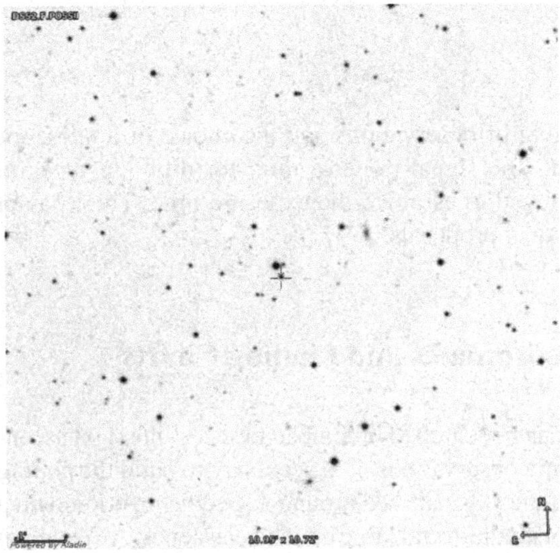

Fig. 13.1 Simbad query and Aladin finding chart for the object 3C 454.3

A quick estimation of the declination range accessible at an observatory site can be made using the equation of the elevation h above the horizon discussed in Chap. 2:

$$\sin h = \sin \phi \sin \delta + \cos \phi \cos \delta \cos(HA) \tag{13.1}$$

where ϕ is the latitude if the observatory, δ the object declination, and HA the hour angle. When observing at the meridian the hour angle is zero, then the equation becomes:

$$h = 90° - \phi + \delta \tag{13.2}$$

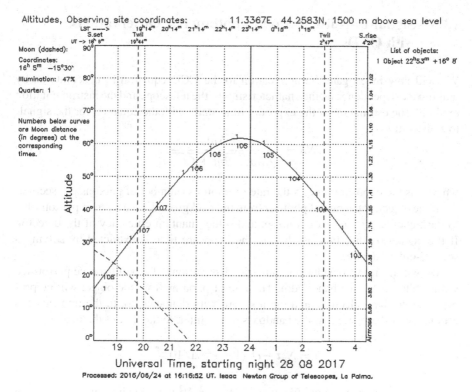

Fig. 13.2 Visibility of 3C 454.3 on the night of 28/8/2016 at a latitude of 44.258°

The range of accessible declinations is defined by $\delta \geq \phi - 90°$. For example, assuming an average latitude of about 45° for Italy, only declinations above $-45°$ can be observed. The additional requirement for optical observations of observing with an air mass smaller than 2, i.e., when objects have an altitude above the horizon larger than 30°, restricts the range of practical observable declinations. For Italy, the effective limit is $-15°$.

Astronomical observations are governed by the times of *sunrise* and *sunset*. The *astronomical darkness* required for astronomical observations corresponds to a Sun position that is more than 18° below the horizon.

For observable objects, the epoch of observation can be approximately estimated by considering that the local sidereal time advances by 2 hours per month and has a value of 0h at midnight on 21 September. The checking of the observability of an object is greatly simplified by *Staralt* software[1] that shows the altitude above the horizon during the whole night (Fig. 13.2).

---

[1] http://catserver.ing.iac.es/staralt/.

13.3 Signal-to-Noise Ratio in Photometric Observations with CCDs

We will now investigate the application to the CCD equation to observations by making the dependence on the characteristics of the telescope and the instrumentation explicit. The estimation of the exposure time starts with the equation for the signal-to-noise ratio [7]:

$$SNR = \frac{S\eta t}{\sqrt{S\eta t + B\eta t + Ct + R^2}} \tag{13.3}$$

where t is the exposure time, S the rate of photons (number of photons per second) from the object, B the rate of photons from the background, C the rate of photons due to dark counts, R^2 the readout noise, and η the quantum efficiency of the detector. If the signal-to-noise ratio is chosen, the exposure time is extracted by solving a second-order equation.

We will initially consider photometric observations of a point source performed with a filter of known bandwidth. The rate of photons from the object will be proportional to the collecting area (an aperture with diameter D), to the transmission efficiency of the system, to the bandpass of the filter, to the flux of the object [7]:

$$S = N_p \tau \frac{\pi}{4} D^2 (1 - \varepsilon^2) \Delta\lambda 10^{-\frac{m}{2.5}} \tag{13.4}$$

where τ is the transmission efficiency, $\frac{\pi}{4}(1 - \varepsilon^2)D^2$ the effective collecting area of a telescope with diameter D with an obscuration factor ε, $\Delta\lambda$ the filter bandpass, m the magnitude of the object, and N_p is a factor to convert magnitude to flux units, whose value is $10^4 \frac{photons}{scm^2 nm}$ for a zero magnitude star at 550 nm.

The contribution of the sky background is estimated in the region defined by the object, an area of the order of θ^2, with a size equal to the FWHM of the seeing disk or of the diffraction disk. The sky background at Paranal in the UBVRI bands is reported in Table 13.1 for reference.

If the sky background has a magnitude m_B measured in mag/arcsec^2, the corresponding photon rate is [7]:

Table 13.1 Optical sky brightness at Paranal [6]

Band	Mean sky brightness (mag arcsec^{-2})
U	22.28
B	22.64
V	21.61
R	20.87
I	19.71

$$B = N_p \tau \frac{\pi}{4} D^2 (1 - \varepsilon^2) \Delta\lambda 10^{-\frac{m_B}{2.5}} \theta^2 \qquad (13.5)$$

The estimation of the signal of extended sources is identical to the estimation of the sky background contribution. There are two different regimes of signal-to-noise ratio. When the object contribution is dominant, i.e., for bright sources, $S \gg B$, we have the *photon noise limited regime*:

$$SNR = \sqrt{S\eta t} \propto D\sqrt{t} \qquad (13.6)$$

where the exposure time is proportional to the reciprocal of the squared telescope diameter, D^{-2}. When the background is dominating, $B \gg S$, we have the *background limited regime*:

$$SNR = \frac{S}{\sqrt{B}} \sqrt{\eta} \sqrt{t} \propto D\theta^{-1}\sqrt{t} \qquad (13.7)$$

where the exposure time is still proportional to the reciprocal of the squared telescope diameter, but also to the size of the object image, as $\frac{\theta^2}{D^2}$. The relation between the time and the telescope aperture underlines the necessity of telescopes with large collecting areas for observations in either regime. But in the background limited regime, the dimension of the star image is as important as the telescope aperture, pushing to building facilities at sites with a good seeing, that offer also the additional advantage of a smaller sky background. When the instrumentation is operating in the *background and diffraction limited regimes*, the image size is given by the Airy disk and is proportional to the reciprocal of the telescope diameter D^{-1}; thus, the exposure time shows a steeper dependence on the aperture size, behaving as D^{-4}. Generally, the exposure time is proportional to the square of the signal-to-noise ratio: moving from a signal-to-noise ratio of 5 to a ratio of 50 requires an exposure time longer by two orders of magnitude. The signal-to-noise ratio increases as the square root of the source flux in the photon noise limited regime and as the source flux in the background limited regime.

The signal-to-noise ratio achievable with a fixed exposure time of 1800 s in different conditions is shown in Fig. 13.3. The performances that can be achieved with three ground-based telescopes with diameters of 4, 8, and 10 m are compared with the performance of the Hubble Space Telescope, with a diameter of 2.4 m. The image size used for the estimation is 1.0 arc sec for ground-based telescopes and 0.1 arc sec for HST [7]. The background noise is 22 mag arc sec^{-2} and 23 mag arc sec^{-2} for ground-based telescopes and for HST, respectively [7]. The parameters of the detector are: $\eta = 0.80$, readout noise of 5 electrons/pixel, dark current 0.005 electrons/pixel/s, and total transmittance of the system 0.3 (see also [7]).

Space-based telescopes are able to access fainter magnitudes than ground-based telescopes, due to the operation at the diffraction limit and the smaller background.

The planning of photometric observations in the near infrared has to tackle a large background (Table 13.2).

Fig. 13.3 Signal-to-noise ratio versus magnitude for ground-based telescopes with diameters of 4, 8, and 10 m and for the Hubble Space Telescope with an exposure time of 1800 s

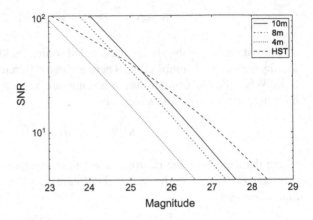

Table 13.2 Infrared sky brightness at Paranal, measured in units of magnitude per square arcsec [1]

Band	Mean sky brightness (mag arcsec^{-2})
J	16.5
H	14.4
L	3.9

13.4 Signal-to-Noise Ratio in Spectroscopic Observations with CCDs

The signal-to-noise ratio in the observation of point-like sources with slit spectrometers is given by the familiar equation [7]:

$$SNR = \frac{St}{\sqrt{St + Bt + Ct + R^2}}, \tag{13.8}$$

where t is the exposure time, and S, B, C are now the rates of photons per pixel. However, in spectroscopic observations, one axis of the spectrum frame is the dispersion axis. The quantity $\Delta\lambda$ is now the product of the plate factor and of the slit width (projected on the detector).

13.5 At the Observatory

A guideline to the operations to be performed at optical telescopes is given in [5]. At the arrival at the observatory, the observer should check that the required instrumentation is mounted on the telescope. The observer should have a working checklist with the names of the objects, the coordinates, the finding charts, the required setup (photometric filters or spectrometer dispersion), and a list of standard objects.

The instrumentation is operated by the personnel of the observatory and not by the observers themselves, generally. The night schedule of the observer should take into account the overhead times required for pointing the telescope, changing the setup and for the storage of the secured data. Additional time is required to observe the calibration stars and to acquire the bias and flat frames. The group of standard stars should span a range of zenith distances, i.e., of air masses, that includes the interval of air masses of the target objects, and a range of magnitudes and colors that are close to those of the targets. The standard stars must be observed regularly during the night, mixing them with the observation of the science targets, to have a wide coverage of zenith distances, that can be used to estimate the extinction.

The absolute spectrophotometry requires the choice of a slit wide enough to avoid light losses. The slit should be aligned along the parallactic angle, the direction of atmospheric dispersion, unless the effect of differential atmospheric refraction is small. Since the spectral resolution depends on the slit width, sometimes a partial light loss is considered acceptable; a wider slit is used only for standard stars. For each acquired spectrum, a spectrum of the lamp must should be secured.

13.6   Radio Observations

The signal-to-noise ratio for radio astronomical observations with single instruments is given by the radiometer equation [9]:

$$SNR = \frac{T_{obj}}{T_{rms}} = \frac{T_{obj}}{T_{sys}}\sqrt{\Delta\nu\tau} \tag{13.9}$$

where T_{obj}, T_{rms}, T_{sys} are the temperatures of the target object, of the rms noise, and of the system, $\Delta\nu$ is the bandwidth of the instrumentation, and τ is the integration time. The temperature of the target object is the brightness temperature measured by the power collected at the radio telescope, $2k_B T_{obj}$. The system temperature contains both the contribution of the sky and of the receiving system. The system temperature is characterized by the *System Equivalent Flux Density* (SEFD) $\frac{2k_B T_{sys}}{A_e}$.

13.7   Interferometric Observations

The planning of interferometric observations is complex and the reader is referred to the specific observatory tools that have been developed for the astronomical community. Some of them are briefly reported here. The preparation of a proposal at optical interferometers can be performed using *The Astronomical Software to PRepare Observations* (ASPRO).[2] The tool allows the preparation at different facilities,

---

[2]http://www.jmmc.fr/aspro_page.htm.

including the VLTI. The software computes the visibilities of the targets and the projection of the baselines during the observation. The visibility of observations at VLTI can be estimated for the different instruments with a dedicated software tool.[3] The VLBI has an online calculator[4] that allows to select different instruments of the array and computes the sensitivity and the uv coverage. The VLA offers an exposure time calculator for the different configurations.[5]

13.8　Observing Proposals

The issues related to the writing of proposals for observing time can be discussed using the observations at optical telescopes as a guideline. The core of a proposal [2–4, 8] at any astronomical facility is the scientific justification. The level of competition to access the facilities is high: The ratio of the submitted to the accepted proposals ranges can be as high as 5 or 6 at large telescopes and the HST. Since the cost of building and operating the facilities is high, the *Time Allocation Committees* (TAC) request proposals that maximize the science output. Generally, one or two years after they have been collected by the proposer(s), data become public.

The committees of astronomical facilities periodically (two or three times per year) send out calls for the submission of proposals. Time is typically allocated within semesters. The submitted proposals should be written underlining the originality of the project, but in a clear language. The reviewers generally examine and grade tens of proposals and must be able to summarize the main points of each of them in a short time.

Following the sample proposal for optical observations provided by NOAO,[6] the structure of a telescope proposal can be summarized as follows:

- Abstract,
- Scientific Justification,
- Experimental Design/Technical Feasibility,
- Use of Other Facilities,
- Long-term Details,
- Previous Use of the Observatory Facilities,
- Technical Description, and
- Target Table.

A first choice to be made is the modality of observation: *Visitor, Service, Target of Opportunity* (ToO). The Visitor mode is the field trip at the facility, where the proposers(s) interact with the personnel and could take last minute decisions about the observations. The Service mode demands the execution of the observations planned

---

[3]http://www.eso.org/observing/etc/bin/gen/form?INS.NAME=VISCALC+INS.MODE=CFP.

[4]http://www.atnf.csiro.au/vlbi/calculator/.

[5]https://obs.vla.nrao.edu/ect/.

[6]https://www.noao.edu/noaoprop/help/sample.pdf.

by the remote observer(s) to the personnel on site. This mode is very suited for programs that do not require any sudden change from the planned schedule. The Target of Opportunity mode is devoted to not predictable astronomical events or the follow-ups of observation in other parts of the spectrum. The observations of ToO targets can be executed during the nights assigned to other programs, interrupting the ongoing observations, with a time slot of one to two hours.

The *Abstract* is a key section that must clearly and concisely summarize the main points of the proposed research and their justification not only for specialists, but also for non-specialist members of the committee.

The *Scientific Justification* should explain why the proposed observations are interesting, the scientific questions to which they will provide the answer, and their impact on astronomy. The fact that an object has not been observed before does not justify, in itself, the awarding of telescope time. If the proposed observations will be compared with models, it is necessary to describe the possible constraints that will be provided by the observations.

The *Experimental Design/Technical Feasibility* section is the roadmap describing the details of the observations. The proposer(s) should explain why and how the proposed observations must be performed at the chosen facility with the chosen instrument(s), justifying that they are not feasible at smaller telescopes. The criteria for the selection of targets must be discussed in detail, in particular the sample size, outlining the minimum number of objects required to address the science problem.

The section *Use of Other Facilities* discusses how the observations of the proposal are linked to observations at other facilities, on ground or in space. The proposers(s) should show that the existing data are not sufficient to test the proposed question. The data to be collected could be the support to previous work or be the precursor of data to be secured at larger facilities. Again, the observations must be framed in the big picture. Existing work in support of the proposed plan should be mentioned.

The section *Long-Term Details* is reserved to observation programs extending for more than one semester and should describe the instrumentation and the number of nights needed to complete the project.

The section *Previous Use of the Observatory Facilities* describes the previous allocation of time at the observatory and the update of the project status. The ability to publish relevant publications within a short time is a major point for proposers, in particular if the previous allocated time at the facility has produced publications. The existence of observatories is supported by the flow of publications produced by the users of their instruments, and the most part of observing time should lead to publication in a short time.

The sections called *Technical Description* and *Target Table* are the summary of the observing details: the number of nights at each instrument, the acceptable months within the semester, and possible scheduling problems. For optical observations, the nights are more or less dark, according to the fraction of the Moon illumination, i.e., the fraction of the lunar disk that is illuminated at local civil midnight, where 1.0 corresponds to full illumination. The *dark time* corresponds to an illumination fraction smaller than 0.4, while the *gray time* corresponds to an illumination fraction in the range from 0.4 to 0.7.[7] The request of the precious dark nights must be justified, explaining in detail if it is mandatory due to the faintness of the sources and/or the high resolution requested.

For each target, the astronomical coordinates, the epoch, the magnitude, the photometric filter, and/or the spectrometer configuration must be provided, together with the exposure time. The exposure time estimated with the procedures described is accompanied by several overheads that must be included to assess the effective duration of an observation.

Problem

13.1 The geographical coordinates of the Loiano Observatory in Italy are $11°20'12''$ North and $44°15'30''$ East. Discuss whether the following objects are observable and determine the period of observability during the year:

- R CrB
- OJ 287
- η Carinae
- 3C 454.3
- NGC 1275
- The Large Magellan Cloud

References

1. Cuby, J. G., Lidman, C., Moutou, C.: The Messenger **101**, 3 (2000)
2. Duchene, G., Duvert, G.: Preparation of VLTI Observations. NewAR **51**, 650 (2007
3. Fomalont, E.B.: Preparing a competititve radio proposal. http://adsabs.harvard.edu/abs/2005xrrc.procE6.03F
4. Kervella, P., Garcia, P. J. V.: Preparing an ESO proposal. NewAR **51**, 658 (2007)
5. Massey, P., Jacoby, G. H.: CCD Data: The Good, The Bad, and The Ugly. In: Astronomical CCD observing and reduction techniques, edited by Steve B. Howell. ASPC **23**, 240 (1992)
6. Patat, F.: UBVRI Night Sky Brightness at ESO-Paranal during sunspot maximum. The Messenger **no115**, 18 (2004)
7. Schroeder, D.: Astronomical Optics. Academic Press (1999)
8. Walsh, J.: On HST Proposal Writing. http://www.spacetelescope.org/static/archives/stecfnewsletters/pdf/hst_stecf_0034.pdf
9. Wilson, T. L., Rohlfs, K. and Hüttemeister, S.: Tools of Radio Astronomy, Springer (2009)

---

[7]https://www.eso.org/sci/observing/phase2/ObsConditions.html.

Chapter 14
After Observation: Data Analysis

This chapter describes the techniques of data reduction and analysis for different astronomical observations. A short primer in statistics and signal processing is given. Then, the techniques used for optical photometric and spectroscopic data, radio data, and interferometric data are presented. The observer should reduce data as soon as possible, starting just after the end of the observing night/run, to check the possible problems in the observations and to have a preliminary set of results. The variety of instruments described so far require different analysis pipelines that will be discussed in detail for each region of the spectrum.

14.1   A Primer in Astronomical Statistics

The astronomical statistics is discussed in detail by [11]. The first exploration of collected data is the search for possible correlations between the measured quantities and other properties [11]. An historical example is the Hubble diagram reporting the velocity against the distance for 24 galaxies (Fig. 14.1). The data suggest a possible linear relation between the two quantities. The correlation can be estimated by the Pearson correlation coefficient or by the nonparametric Spearman rank correlation coefficient [11].

The presence of a correlation between two variables does not necessarily imply the existence of a physical relation between them. Artificial correlations can appear, generally due to the criteria of the sample selection or, sometimes, due to the relation of the two quantities to a third one. The *selection effects* are a potential source of problems in several astronomical contexts. The diagram of the luminosity of radio sources as a function of the distance modulus seems to suggest that the most distant objects are brighter (Fig. 14.2), as discussed by [9].

The sample of the diagram is a *flux limited sample*, whose elements are selected by the criterion of being brighter than a fixed threshold. Faint sources are not observed at

© Springer International Publishing Switzerland 2017   165
R. Poggiani, *Optical, Infrared and Radio Astronomy*,
UNITEXT for Physics, DOI 10.1007/978-3-319-44732-2_14

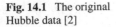

Fig. 14.1 The original
Hubble data [2]

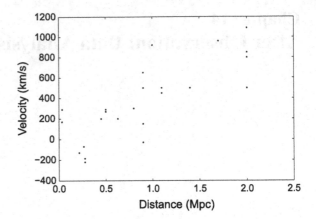

Fig. 14.2 Luminosity of
radio sources as a function of
the distance modulus; data
from [9]

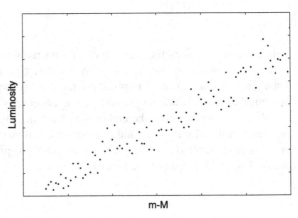

large distances because they are below the threshold. Bright sources are not observed
at small distances because they are rarer than faint sources. The minimum luminosity
L_{min} is related to the flux limit S_{min} by:

$$L_{min} = 4\pi d^2 S_{min} \qquad (14.1)$$

where d is the distance of the object; the flux limit appears as a straight line with slope
2. The correlation between the luminosity and the distance is an apparent correlation
due a selection effect called *Malmquist bias*. Often the luminosities L_1, L_2 in two
different regions of the spectrum are investigated. If the samples are selected if they
are brighter than the flux thresholds in the two domains, the two luminosities will
show an apparent correlation with slope 1, since they are both correlated with the
distance. The presence of an effective correlation can be checked by investigating
the relation between the fluxes S_1, S_2.

   If a genuine correlation is found, a model curve is fitted to the data and the quality
of the fit evaluated. The methods of least squares and the more general maximum

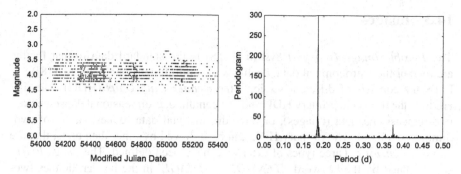

Fig. 14.3 Time series of the variable star δ Cephei (*left*) and Lomb–Scargle periodogram (*right*); data from AFOEV, http://cdsarc.u-strasbg.fr/afoev/

likelihood method are of common use. A detailed discussion has been presented by [11].

14.2 Time Series Analysis

The analysis of time series in astronomy spans a wide range of physical processes [11]. Periodicities are observed in systems with intrinsic pulsations, such as the Cepheids or the RR Lyrae stars, but also in binary systems due to the orbital motion. Transient sources, such as novae, supernovae, and gamma-ray bursts, produce flares. In addition, accretion in binary systems, such as the X-ray binaries and the cataclysmic variables, is a stochastic process. The astronomical time series are usually gapped, due to weather conditions or lack of telescope time. The standard Fourier techniques are replaced by the *Lomb–Scargle periodogram* [4, 10], the equivalent of the power spectrum. If the data points h_i are measured at the times t_i, the Lomb normalized periodogram is:

$$P_N(\omega) = \frac{1}{2\sigma^2} \left[\frac{\sum_i (h_i - \bar{h}) \cos \omega(t_i - \tau))^2}{\sum_i \cos^2 \omega(t_i - \tau)} + \frac{\sum_i (h_i - \bar{h}) \sin \omega(t_i - \tau))^2}{\sum_i \sin^2 \omega(t_i - \tau)} \right]$$

$$(14.2)$$

where $\bar{h}$ and σ^2 are the mean and the variance of the data and $\tan(2\omega\tau) . = \frac{\sum_i \sin 2\omega t_i}{\sum_i \cos 2\omega t_i}$. The periodogram is equivalent to the estimation of the harmonic content of the time series using a least square fit to the function $A \cos \omega t + B \sin \omega t$.

An example of the application of the Lomb–Scargle periodogram is shown in Fig. 14.3.

14.3   Images

The *Flexible Image Transport System* (FITS) is the standard data format for the archival of the astronomical data, dealing with multidimensional data (see [8]).[1] A FITS file consists of different sections, the *Header Data Units* (HDU). The first HDU of the file is the primary HDU and can contain one-dimensional data (spectra), two-dimensional data (images), or three-dimensional data (cubes, stacks of two-dimensional data). The primary HDU can be followed by other data units, that are named *extensions*. Three types of extensions have been defined. The *Image Extension*, defined by the keyword *XTENSION* = *'IMAGE'* in the header, defines two-dimensional arrays of data. The *Table Extension*, labeled by the keyword *XTENSION* = *'TABLE'* is standard data tables stored in a readable ASCII format. The third extension, the *Binary Table Extension*, is marked by *XTENSION* = *'BINTABLE'* in the header and stores data tables in binary format. Binary tables are accessed more easily and demand a smaller disk space.

The Header Units comprise an ASCII *Header Unit* and a *Data Unit*. The header consists of a set of records with a length of 80 characters each; each record contains an eight character keyword and its value, *KEYWORD* = *VALUE*. The minimal FITS header contains the information about the dimension and the format of the data:

```
SIMPLE  =                       T /FITS header
BITPIX  =                      16 /No.Bits per pixel
NAXIS   =                       2 /No.dimensions
NAXIS1  =                     530 /Length X axis
NAXIS2  =                     530 /Length Y axis
END
```

The BITPIX keyword defines the number of bits per pixel, while NAXIS describes the dimension of the data, an array with NAXIS1×NAXIS2 size. Other keywords that can be found in the header are related to the date and UT time of observation (DATE-OBS, UT), the telescope and the instrument (TELESCOP, INSTRUME), the observer (OBSERVER), the object name (OBJECT), the exposure time (EXPTIME), the air mass (AIRMASS), and possible comments (COMMENT). The header must be mandatorily closed with the keyword END. The data unit contains data stored in a format defined by the keyword BITPIX. The supported data types for arrays are: 8-bit integers, 16-bit signed integers, 32-bit signed integers, 32-bit single-precision floating real numbers, and 64-bit double-precision floating real numbers.

The FITS format has been adapted for the interferometric observations. The standards for optical and radio interferometric are OIFITS [7],[2] UVFITS, and FITS-IDI.[3]

---

[1] http://fits.gsfc.nasa.gov/fits_home.html.

[2] http://fits.gsfc.nasa.gov/registry/oifits.html.

[3] http://fits.gsfc.nasa.gov/registry/fitsidi.html.

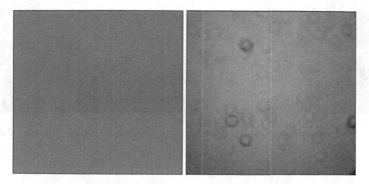

Fig. 14.4 Photometric observations with CCDs: bias frame (*left*) and flat frame (*right*)

14.4 Reduction of Photometric Observations

During an optical photometric observing session, several frames are secured (Fig. 14.4) [5, 8]:

- image of the science target,
- bias frames,
- flat field frames for each photometric filter, and
- image of photometric standard stars.

The bias frames are combined together to build the master bias by using the median pixel by pixel. The flat field frames of each filter are combined pixel by pixel using the median to build the master flat for that filter. The flat is normalized by dividing it by its average value to produce a normalized response pixel by pixel. Finally, the bias and the flat frames are applied to the science frame following Eq. 6.6.

14.5 Reduction of Spectroscopic Observations

We will discuss slit spectroscopy of point-like sources, such as stars. During the observing run, different frames must be secured [5, 6]:

- Spectrum or spectra of the science target
- Bias frames
- Flat frames for each configuration of the dispersing elements
- Spectra of lamps for wavelength calibration
- Spectra of spectrophotometric standard stars.

The raw spectrum, the bias spectrum, the flat spectrum, and the lamp spectrum are shown in Fig. 14.5.

The preparation of the master bias frame and the master flat frame follows the same guidelines of the photometric reduction. The master bias and the master flat frames

Fig. 14.5 Spectroscopic observations with CCDs: raw spectrum, bias frame from *left* to *right*, flat frame, lamp frame

are the median combinations of the set of bias and flat frames, respectively. The peculiarity of the spectroscopic flat frames will be discussed below. The spectrum frame must in principle be reduced with the master bias and the master flat.

The dispersion axis of the science frame in Fig. 14.5 is along the vertical direction. The spectrum is contained in the strip extending along the whole frame. The dispersion in pixel must be calibrated in wavelength using the lamp spectrum. The spatial axis is along the horizontal direction. The spectrum extends along a few pixels. Emission and absorption lines appear as button like or thinned regions of the spectrum strip. The horizontal lines spanning the whole frame are the emission lines of the sky. The extraction of the spectrum of a target requires the identification of the spectrum, the definition of an aperture for the extraction and the identification of a region of sky background. A cross section of the spectrum along the spatial direction shows a peaked curve, flanked by the flatter regions of the background (Fig. 14.6). The spectrum aperture is chosen to completely contain the spectrum. The spectrum of the science target can be extracted, for each point on the dispersion axis, by summing the content of the pixels in the object aperture and subtracting the value of the background; the weighting factor of the pixels is one. In the *optimal extraction* method [1], the weighting factor is the variance of the pixel intensities. The sky background spectrum is extracted using an aperture or two symmetrical apertures close to the spectrum aperture, carefully avoiding possible spectra of unwanted objects. The position of the centroid of the spectrum along the spatial axis is obtained by fitting the profile. The contribution of the background is subtracted from the spectrum.

The spectroscopic flat frames, needed to correct for the different pixel sensitivities, are secured using a lamp with a featureless spectrum and do not show the almost uniform intensity distribution typical of the photometric flat frames. A one-dimensional flat response profile is produced by averaging along the array rows (Fig. 14.7). The shape of the lamp spectrum is removed by fitting a low-order curve to the flat response profile. The flat field is divided by the fitted curve, leaving only the variation in sensitivity at small scales.

The extracted spectrum so far is still a function of the pixel number along the dispersion direction that must be transformed into a wavelength scale. The *wavelength calibration* is performed using the spectrum of the calibration lamp that shows a comb of narrow lines at known wavelengths (Fig. 14.8). The lamp is physically different from the featureless lamps used for spectroscopic flat fielding. Typical lamps

Fig. 14.6 Section of a spectrum along the spatial direction

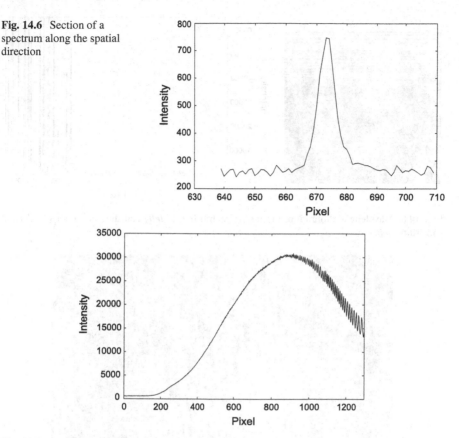

Fig. 14.7 Spectroscopic flat frame: profile of flat response

are HeNeAr, ThAr, FeAr, and CuAr.[4] The lamp spectrum must be secured for each science target. The user must identify some features at known wavelength in the lamp spectrum, preferably distributed over the whole spectrum. A low-order polynomial relates the wavelength to pixel number, the *dispersion solution*. The dispersion solution is applied to the spectrum of the science frame to produce a wavelength calibrated spectrum.

The next step is the choice between the *flux calibration* and the *spectrum normalization*. The flux calibration requires the observation of spectrophotometric standard stars; their spectrum is extracted with the same strategy of the target spectrum. The spectrum of the standard star is compared to an archived spectrum of the same object. The conversion factor is the ratio of the observed to the archived spectrum. The flux calibrated spectrum is then corrected for the atmospheric extinction. The standard star should be observed at an air mass similar to that of the science target. The procedure of spectrum normalization removes the instrumental profile from the spectrum,

---

[4]A spectral atlas of lamps is available at http://iraf.noao.edu/specatlas/.

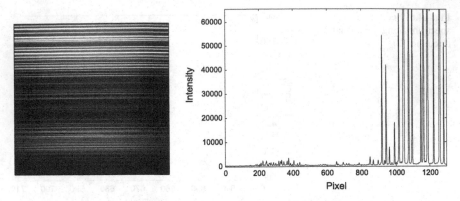

Fig. 14.8 Wavelength calibration: the lamp spectral frame (*left*) and the one-dimensional lamp spectrum (*right*)

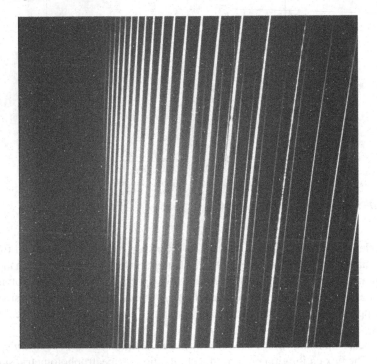

Fig. 14.9 Raw echelle spectrum

flattening the continuum. The normalization is performed by fitting a polynomial to the spectrum and dividing the spectrum by the fitted curve.

The high resolution cross dispersed spectra secured with echelle spectrographs appear as a set of short spectra related to different orders arranged as in a waterfall order. There are several distinct aperture and dispersion axes (Fig. 14.9). The same pattern is observed in the echelle flat frames.

The bias frames and the flat frames are combined following the same recipes described above for slit spectroscopy. The flat field is normalized, ensuring that all apertures. i.e., all orders, are correctly identified. A cut along the dispersion axis of each order will show the spectral shape of the lamp. The flat is then normalized. Finally, the bias and the flat frame corrections are applied to the science frame. A cut of the corrected frame orthogonal to the dispersion axis will show an almost flat baseline with peaked functions corresponding to the echelle orders. The rest of the analysis is performed on the same lines of slit spectroscopy, applying the techniques of slit spectroscopy to each order, to provide a wavelength calibrated spectrum. Flux calibration of echelle spectra is more difficult than for slit spectra, since they are split into short sections.

14.6 Reduction of Infrared Data

The large background of near infrared observations [3, 5, 6] prevents the integration of signals with long exposure times, as in optical astronomy. The integration times do not exceed a few minutes and can become smaller than one second in the thermal infrared region. The signal in the pixel j is given by:

$$I_j = (S_j + B_j)F_j + D_j \qquad (14.3)$$

where S_j, B_j, D_j are the signals of the target source, of the background, and of the dark current and F_j is the flat field response. The background can be orders of magnitude larger than the target signal and must be precisely measured before its subtraction. If the field is not excessively crowded, a background frame can be produced by successive observations of the region around the source by slightly moving the telescope at each exposure. The dithered images are combined together by computing the median pixel by pixel to reject the sources.

The signal in the background frame is:

$$I_j^{background} = B_j F_j + D_j \qquad (14.4)$$

The subtraction of the background from the science frame removes the contribution of the dark current. The signal of the source is extracted from the difference by normalizing with the flat field:

$$S_j = \frac{I_j - I_j^{background}}{F_j} \qquad (14.5)$$

The reduction in the observations in the mid-infrared is more complicate that in the near infrared. Each observation comprises four images: two frames are secured for two chopping positions of the secondary at each of the nodded positions in the sky. The reduction procedure is a sequence of subtraction steps. The frames secured at the same nodded position are subtracted to remove most of the background from the sky and the telescope. The rest of the background is removed by subtracting the chop-subtracted image at one of the nodding positions.

14.7   Reduction of Radio Data

The radio observations must be corrected for the effects of the atmosphere [3, 5, 6]. The calibration of data in the centimeter to meter region is performed by injecting in the receiver system pulsed signals with known amplitude produced by noise diodes. The calibration procedure secures data without and with the pulsed signal. The receiver noise is estimated using Eq. 9.18. The millimeter and submillimeter range is more sensitive to the effect of the atmosphere and to its variability. The calibration process adopt a chopping techniques as in infrared astronomy, measuring the receiver noise (Eq. 9.18) and the output of the observation of a blank sky area. The output of the system when a resistive load is measured is proportional to the sum of the load temperature T_{ref} (usually the room temperature) and of the receiver noise T_r:

$$output_{ref} \propto T_{ref} + T_r \qquad (14.6)$$

while the output measured when pointing at the sky is proportional to the sum of the receiver noise and of the contribution of the sky at temperature T_{sky} and of the contribution of the ground at temperature T_g:

$$output_{sky} \propto T_r + F T_{sky} + (1 - F) T_g \qquad (14.7)$$

where F is the forward efficiency, the fraction of power in the forward beam of the system. The sky temperature is related to the atmosphere temperature T_{atm} and to the optical depth τ, according to $T_{sky} = T_{atm}(1 - e^{-\tau})$. Subtracting Eq. 14.7 from Eq. 14.6:

$$output_{cal} = output_{ref} - output_{sky} \propto F T_{ref} e^{-\tau} \qquad (14.8)$$

The value of the optical depth is used to correct the signal from the astrophysical source, according to:

$$output_{source} \propto T_{a,a} e^{-\tau} \qquad (14.9)$$

14.8   Reduction of Interferometric Data

The interferometric observations are performed by alternating the observation of the science target and of a precisely known calibrator [3, 5, 6]. The evolution of the amplitude of the fringes is a square wave curve, in principle. The fluctuations of the atmospheric properties and of the electronic gain of the instruments can affect the amplitude of the square wave. The effect of the drifting amplitude is removed by fitting the part of the curve corresponding to the calibrator data with a low-order polynomial and applying the correction to the science target data.

14.9   Astronomical Software

The *Astrophysics Source Code Library* (ASCL)[5] is an online repository for software relevant for astronomers, including software developed for published papers. The archive is continuously updated with new source codes.

Some software packages are widely used and are briefly summarize here.

- The *SAOImage DS9*,[6] is an astronomical imaging and data visualization application.
- The *Image Reduction and Analysis Facility* (IRAF)[7]: developed at the National Optical Astronomy Observatories offers several programs for the reduction of optical and infrared astronomy data. Several observatories, such as the HST, have produced packages that can be added to the standard IRAF package. IRAF includes also the scripting capability.
- The commercial software *Interactive Data Language* (IDL).[8] The IDL Astronomy User's Library http://idlastro.gsfc.nasa.gov/ is an online archive of astronomical software written in IDL and the starting point for other resources.
- The *Astronomical Image Processing System* (AIPS)[9] is a package dedicated to the calibration and analysis of radio interferometric data.
- The *Common Astronomy Software Applications* (CASA)[10] is a toll for the processing of data from single dishes and radio interferometers.

---

[5]http://ascl.net/.

[6]http://ds9.si.edu/site/Home.html.

[7]http://iraf.noao.edu/.

[8]An open source IDL compiler is available at http://gnudatalanguage.sourceforge.net/.

[9]http://www.aips.nrao.edu/.

[10]http://casa.nrao.edu/index.shtml.

Problems

14.1 Describe the steps required to reduce photometric frames in observations with CCDs.

14.2 Describe the steps needed to reduce the spectroscopic frames in CCD observations.

14.3 Discuss the purpose of the calibration lamp in spectroscopic observations with CCDs.

References

1. Horne, K.: An Optimal Extraction Algorithm for CCD Spectroscopy. PASP 98, 609 (1986)
2. Hubble, E.: A Relation between Distance and Radial Velocity among Extra-Galactic Nebulae. PNAS **15**, 168 (1929)
3. Lèna, P. et al.: Observational Astrophysics. Springer-Verlag Berlin Heidelberg (2012)
4. Lomb, N.R.: Least-squares frequency analysis of unequally spaced data. Ap&SS **39**, 447 (1976)
5. Oswalt, T. D., McLean, I. S.: Planets, Stars and Stellar Systems. Volume I: Telescopes and Instrumentation. Springer (2013)
6. Oswalt, T. D., Bond, H. E.: Planets, Stars and Stellar Systems. Volume II: Astronomical Techniques, Software, and Data. Springer (2013)
7. Pauls, T. A., Young, J. S., Cotton, W. D., Monnier, J. D.: A Data Exchange Standard for Optical (Visible/IR) Interferometry. PASP **117**, 1255 (2005)
8. Pence1, W. D., Chiappetti, L., Page, C. G., Shaw, R. A., Stobie, E.: Definition of the Flexible Image Transport System (FITS), version 3.0. A&A **524**, A42 (2010)
9. Sandage, A.: The Redshift-Distance Relation. 111. Photometry and the Hubble Diagram for Radio Sources and the Possible Turn-On Time for QSOS. ApJ **178**, 25 (1972)
10. Scargle, J.D.: Studies in astronomical time series analysis. II - Statistical apsects of spectral analysis of unevenly spaced data. ApJ **263**, 835 (1982)
11. Wall, J. V. and Jenkins, C. R.: Practical Statistics for Astronomers. Cambridge University Press (2003)

Index

A
Aberrations, 35
Absolute magnitude, 8
Acousto-optical spectrometer, 133
Active optics, 44
Adaptive optics, 54
Air mass, 85
Airy pattern, 48
Aladin, 24
Alt-azimuth mount, 34
Alt-azimuth system, 20
Antenna, 121
Antenna temperature, 124, 129
Aperture synthesis, 142
Apparent magnitude, 8
Astigmatism, 37
Astronomical spectroscopy, 91
Astrophysics Data System, 28
Atmosphere, 9, 52
Autocorrelation spectrometer, 132

B
Background, 15
Bias frame, 81, 101
Blazed grating, 97
Bolometer, 113
Bolometric magnitude, 8, 79

C
Cassegrain telescope, 38
Charge-Coupled Devices (CCDs), 69
Chopping, 111
CMB, 134
Coherence, 139
Coma, 37

Correlation, 165
Correlation receiver, 130

D
Dark current, 70
Declination, 20
Deformable mirror, 54
Diffraction, 45
Distortion, 37

E
Echelle, 97
Electromagnetic spectrum, 4
Epoch, 22
Equatorial mount, 34
Equatorial system, 20
Exposure time, 158
Extinction, 85
Extrinsic photoconductor, 112

F
Fabry-Perot, 102
FITS, 167
Flat frames, 81
Fourier Transform Spectrometer, 103
Fried parameter, 53

G
Grating, 95
Gregorian telescope, 38
Grism, 97
Ground-based Astronomy, 13

© Springer International Publishing Switzerland 2017
R. Poggiani, *Optical, Infrared and Radio Astronomy*,
UNITEXT for Physics, DOI 10.1007/978-3-319-44732-2

Printed in the United States
By Bookmasters